SMART PRACTICE

How to
Study
Mathematics

How to Study Mathematics

Effective Study Strategies for College and University Students

Peter Schiavone
University of Alberta

Canadian Cataloguing in Publication Data

Schiavone, Peter, 1961-
    How to study mathematics: effective study strategies for college and university student

(Small practices)
Includes index.
ISBN 0-13-906108-8

1. Mathematics - Study and teaching (Higher).   I. Title.   II. Series.
QA11.S34 1998   510'.71'1   C97-931595-6

Prentice-Hall, Inc., Upper Saddle River, New Jersey
Prentice-Hall International (UK) Limited, London
Prentice-Hall of Australia, Pty. Limited, Sydney
Prentice-Hall Hispanoamericana, S.A., Mexico City
Prentice-Hall of India Private Limited, New Delhi
Prentice-Hall of Japan, Inc., Tokyo
Simon & Schuster Southeast Asia Private Limited, Singapore
Editora Prentice-Hall do Brasil, Ltda., Rio de Janeiro

ISBN 0-13-906108-8

Publisher: Pat Ferrier
Marketing Manager: Tracey Hawken
Editorial Assistant: Melanie Meharchand
Production Editor: Amber Wallace
Production Coordinator: Jane Schell
Cover Design: David Cheung

1 2 3 4 5   WC   02 01 00 99 98

Printed and bound in Canada.

Visit the Prentice Hall Canada Web site! Send us your comments, browse our catalogues, and more at **www.phcanada.com**. Or reach us through e-mail at **phcinfo_pubcanada@prenhall.com**.

To Linda and Francesca

Contents

Preface

Why This Book was Written

Why do people have so much trouble in mathematics courses? My response to this question is that they *needn't*. There is a proven recipe for success in mathematics courses - and it works for *all* mathematics courses. This book will show you how.

Over the last ten years, as a college and university mathematics instructor, I have taught a wide range of mathematics courses to both traditional and non-traditional students of engineering, science and business. In addition, as Director of a mathematics resource centre (a unit created specifically to help students make the transition to college/university-level mathematics), I have had the opportunity to work closely with students in developing effective study strategies for a wide range of mathematics courses: from algebra, geometry, trigonometry and calculus, to differential equations and statistics. In that time, it has become clear that the most successful students follow a set of key strategies specific to mathematics. These strategies, once regarded as standard in any basic high-school mathematics curriculum, have become less and less familiar to today's student. This, perhaps, accounts for the *fear, anxiety* and dissatisfaction associated with many present-day mathematics courses.

This book presents that very collection of the most effective techniques for maximizing performance in mathematics courses. The techniques themselves are sufficiently general to be applicable to *all* mathematics courses (in particular, algebra, geometry, trigonometry and calculus) at *any* level. As such, this book is strongly recommended to senior high school students, traditional college and university students and non-traditional students returning to high school, college or university after perhaps an alternative career or occupation. The ideas and concepts presented here are essential in building a solid base for subsequent study in any area where mathematics courses are required (for example, in science, engineering or business).

The presentation is divided into six main parts and follows the natural progression through a typical mathematics course.

- In Chapter 1, we reveal *the study habits of successful students* - the established key learning strategies for maximizing performance in mathematics courses, as practised by the most effective students.

- In Chapter 2, we discuss techniques for *effective preparation*: how best to identify and practice the necessary prerequisite skills for a particular mathematics course. The objective here is to ensure that you know what your instructor assumes you know *before* the course begins.

- Chapter 3 is concerned with maximizing performance *during* a mathematics course. We discuss: how to make the most of class time (for example, taking notes in class), how to make effective use of the textbook, maximizing performance in assignments and homework, getting help from your instructor and how to find other sources of help.

- In Chapter 4, we present procedures for *effective problem-solving*. In particular, we demonstrate how to write solutions to problems in a clear, concise and methodical manner. We also examine, in depth, procedures for solving *word problems*: from translating English into mathematics to identifying and applying appropriate mathematical techniques.

- Chapter 5 is devoted entirely to the subject of *examinations*. Here, we provide a comprehensive discussion of all aspects of the examination experience. In particular, we concern ourselves with maximizing performance in examinations: from preparing for and writing an examination to dealing with *exam-anxiety*.

- In Chapter 6, we discuss other resources such as *tutors* and *self-study manuals*. We examine the different aspects associated with hiring and making the most effective use of a tutor. Finally, we mention how to choose supplementary materials, such as self-study manuals, for a particular mathematics course.

This book is unique among study guides and books dealing with the general area of study skills. This is attributed to the fact that the material is aimed specifically at students enrolled in *mathematics courses*. Consequently, the presentation makes use of actual mathematics to illustrate ideas and concepts. It should be noted, however, that the level of mathematics is kept to the minimum necessary for the correct presentation of the material.

I would gratefully like to acknowledge the contribution of the following reviewers: Thomas G. Berry, University of Manitoba; Millie Atkinson, Mohawk College and W.A. Cebuhar, Queen's University.

Finally, I would like to thank the many senior high school, college and university students who took time from their hectic schedules to answer the many questions that led to the idea for this book.

Peter Schiavone Ph. D

Chapter 1
Introduction
THE STUDY HABITS OF SUCCESSFUL STUDENTS

Over the last ten years, I have taught many different types of mathematics courses to many different types of students: from courses in algebra and geometry to courses in calculus and differential equations; from students primarily interested in engineering and science to those pursuing a career in business or education; from college and university students to non-traditional students returning to high-school, college or university. At the end of each course, I make it routine to approach the most successful students and ask them to describe their study habits and any special techniques used to achieve that particular level of success. This information is then compared to the corresponding responses from those students that did not perform as well.

The above 'surveys' consistently point to a set of key learning/study strategies as the primary reason why the most successful students are so effective in the many different aspects of a mathematics course (from taking notes in class to writing examinations). Even more significant is the fact that almost every one of these strategies is absent from the study habits of the less successful students!

The following simple example (from line geometry) illustrates just one of these strategies. It should be noted that the subject matter of the example is unimportant and is chosen for illustrative purposes only. The same *principles* apply to *any* area of mathematics.

Example 1.1

Find the equation of the straight line with slope $m = \dfrac{1}{2}$ passing through the point $(2, 1)$.

There is only one correct answer to this problem $(y = \dfrac{x}{2})$ but there are many different ways of arriving at that answer. When solving a problem in mathematics, *how* you solve the problem is just as important (if not more important) than the answer itself. Below are three different versions of the solution offered by three different students, A, B and C.

SOLUTION A

$$y = \frac{x}{2} \qquad ans$$

SOLUTION B

$$y - 1 \;=\; \frac{1}{2}(x - 2)$$
$$y \;=\; \frac{x}{2} \quad ans$$

SOLUTION C

The equation of a straight line with slope m passing through the point (a, b) is given by (see Textbook, page 315)

$$y - b = m(x - a) \tag{1.1}$$

Here, $m = \frac{1}{2}$, $a = 2$ and $b = 1$ (given in the question). Substituting these values into the equation (1.1) leads to

$$y - 1 \;=\; \frac{1}{2}(x - 2)$$
$$y - 1 \;=\; \frac{x}{2} - 1$$
$$y \;=\; \frac{x}{2} \quad ans$$

All three solutions are 'correct' in that they have arrived at the correct answer. The main difference between the three is in *detail* i.e. the amount of thought/thinking that has actually found its way to the paper. *Detail* is important for the following reasons.

- **It's always easier to *review* than *relearn***

When you've spent some time solving a problem, it is always worthwhile writing down the details of *how* you solved the problem. In a few weeks time (e.g. when reviewing for examinations) you will have covered much more material and it is likely that you will have forgotten how to solve this particular problem (you may have used some special trick or detail which may have taken a significant amount of time to develop) - it's always easier to *review* than *relearn* i.e. why waste time re-learning how to solve the problem when you can read a detailed account of how you solved it earlier - when you were *focused* on this particular material? A detailed account of the problem-solving process provides you with a reference to help you solve similar problems when, say, practicing for examinations.

- **Using a step-by-step solution procedure makes it easier to identify errors**

Following a clear, methodical, step-by-step procedure to solve a problem makes it easier to identify mistakes if, for example, you arrive at an incorrect answer. Simply go through each step until you find the error. This is particularly important when answering questions in assignments and examinations.

- **Score more points in an assignment or examination**

In an examination, if you make a slip which leads to the wrong answer but have nevertheless demonstrated a clear, logical (and correct) procedure, it is likely that you will receive the majority of the marks.

- **Organize your thoughts and develop an effective problem-solving procedure**

Detail is what distinguishes a solution from an answer. The former is an *account* of your thoughts i.e. the problem-solving process (which you should always record) while the latter is the *result* of your thoughts. The *process* is more important that the answer since it can be applied to similar problems whereas the answer is relevant only to a specific problem. Effective problem-solving procedures will assist you in all aspects of academia (and in many aspects of everyday life!) - not only in mathematics courses.

Each of these points (and many more) will be discussed, in detail, in subsequent chapters of this book.

From what has been said above, it is clear that Solution C is by far the most effective of the three solutions to Example 1.1. Solution A gives only the answer. Student A may very well understand the method but has not demonstrated or recorded the procedure. The problem has most likely been solved 'the student's head' or elsewhere (e.g. on a scrap piece of paper). Solving a question in this way is always dangerous: a simple slip may easily lead to the wrong answer and, with no recorded explanation, there is no way to trace the error! That aside, the lack of recorded detail in Solution A means that, in the event of an incorrect answer, the person grading the solution has no reason to award anything other than zero. Also, there is no evidence of an effective problem-solving strategy. Complicated problems cannot be solved 'in your head', they require the development of a logical step-by-step procedure. Solution B gives a little more detail but leaves it to the reader to 'second-guess' the student e.g. we are left to deduce which formula is being applied. As in Solution A, there is no record of the problem-solving process, making it difficult for the grader to award 'part-marks'. Also, the lack of any significant detail means that any mistake would be difficult (if not impossible) to trace - the problem would have to be re-solved from the beginning.

With a little practice, Solution C is neither more time-consuming nor more difficult to produce than Solutions A or B. In fact, the most successful students will implement an effective problem-solving strategy *automatically* on tackling *any* problem. This arises as a natural consequence from the consistent and repeated application of a methodical, step-by-step procedure.

Example 1.1 is extremely simple - but the principles mentioned above are crucial - and become more so as the level of complexity of the problem increases. More on this later.

Writing effective solutions to problems is only one of the ten key strategies used by the most successful students in mathematics courses. The remainder are contained in the following list which also provides a reference to where in this book each strategy is examined in detail.

- *At the beginning, know what your instructor assumes you know*

Before the course begins, make sure you know what your instructor *assumes* you know i.e. identify prerequisite skills and make sure they are in *working order* - practice with some 'warm-up' exercises if only to wake up the 'mathematical side' of your brain. Remember, perhaps more than most subjects, mathematics is cumulative: one part usually depends heavily on a knowledge of the previous part (Chapter 2).

- *Make the most of lectures and classroom time*

Lectures and classroom time serve to target and highlight relevant material (for both learning and examination purposes). They are also used by the instructor to demonstrate proper procedures when, for example, solving problems and to indicate the required standard. Use lectures and classroom time as your main source of all information relevant to your course. Make sure you know how to take notes efficiently so as to maximize your effectiveness in the classroom (Chapter 3).

- *Your textbook is a supplement not a replacement*

Textbooks do not replace lectures/classroom time. There may be far too much material in the textbook or conversely insufficient detail on a topic which the instructor deems to be important (and examinable!) . The instructor will highlight and target the most relevant material. Use the textbook as a supplement i.e. as a source of examples, practice problems and as a back-up (Chapter 3).

- *Write effective solutions to all problems*

Whether you are solving assignment/homework, classroom, practice or examination problems, you should develop an organized problem-solving procedure as discussed above (Chapters 3 and 4).

- *Assignments indicate required standard*

Assignment/Homework problems are assigned to allow you to practice *relevant* examples so that you have an idea of the required standard and an indication of what is and what isn't important (in, for example, course examinations) (Chapter 3).

- *Practice! practice! practice!*

In mathematics, you learn by seeing examples and *doing* exercises. There is no substitute! You cannot swim or drive a car using only the written theory - you must actually perform the deed! Similarly, in mathematics, it's easy to read a solution and believe that you know what to do. However, problem-solving requires that you actually perform the solution by yourself - which is entirely different! Practice establishes procedure and allows you to note patterns in solutions so that eventually the procedure becomes automatic (again, as in swimming and driving a car) (Chapters 3 and 4).

- *Ask! ask! ask!*

Never be afraid to ask. Seek out all sources of help including your instructor. Be polite and professional but ask as many questions as required - this is a crucial part of the learning process (Chapters 3 and 6).

- *Examination technique*

Knowing the material does not necessarily translate into success in examinations. Remember that in an examination you are subjected to a **time constraint**. Consequently, you must be as effective as possible. To this end, there are ways to develop *examination technique* (Chapter 5).

- *Rehearse and dress-rehearse examinations*

Old and practice-examinations with access to full solutions are an essential part of exam-preparation - particularly in eliminating *exam-anxiety* (Chapter 5). They afford the possibility of *rehearsing* the actual examination and therefore reducing the *element of surprise* (Chapter 5).

- *Write examinations to demonstrate your abilities*

When answering questions in examinations, explain to the examiner *exactly* what you are doing. Write clearly and neatly to win as many *part-marks* as possible. You cannot claim that you "know what you are doing" unless you write it down. The examiner requires that you **demonstrate** your abilities (Chapter 5).

In the chapters that follow, each of these ten points is be examined in detail. Examples will be provided from different mathematics courses at different levels.

Chapter 2

BEFORE THE COURSE BEGINS - GETTING READY

Mathematics, perhaps more than any other subject, is cumulative. This means that performance in any particular mathematics course depends heavily on certain skills developed in previous mathematics courses. For every mathematics course there is a specific set of prerequisite skills deemed to be essential by your instructor. These skills will depend on the level and content of the mathematics course you are about to take - but they are usually few in number and almost always form only a *small part* of any formal prerequisite course. This makes them easy to review or practice! Consequently, to prepare effectively, it makes sense not to spend hours reviewing *all* previous material but rather to identify and target those particular prerequisite skills. This forms the basis of an established procedure for effective preparation (listed below) - which doesn't take long, in fact, only a few hours but which is known to significantly affect subsequent performance.

Step 1 *Make sure that you know what your instructor assumes you know.*

Talk to your instructor (or someone who has experience of this course) and identify the most essential prerequisite skills for your course. Ask exactly what the instructor thinks/assumes you know at the beginning of the course. For example, one of the first major topics in a beginning calculus course is concerned with the evaluation of the expression

$$f'(x) = \lim_{h \to 0} \frac{f(x+h) - f(x)}{h} \tag{2.1}$$

where f is a given function and h is some constant. The 'calculus component' of (2.1) consists of the limiting process and the application and interpretation of the quantity $f'(x)$, known as the first derivative of the function $f(x)$. The *precalculus* component (i.e. that which the instructor assumes is known to the student) is everything to do with forming, simplifying and manipulating the expression

$$\frac{f(x+h) - f(x)}{h} \tag{2.2}$$

For example, let $f(x) = \sqrt{x+3}$. Then, before any *calculus* (i.e. limiting process and interpretation) can be applied, we must first simplify and *prepare* the expression (2.2) using *precalculus* techniques:

$$\frac{f(x+h) - f(x)}{h} = \frac{\sqrt{x+h+3} - \sqrt{x+3}}{h} \tag{2.3}$$
$$= \frac{1}{\sqrt{x+h+3} + \sqrt{x+3}}$$

6

The expression (2.3) is now in a form suitable for the application of *limit theory* - the *calculus part*. In arriving at (2.3) we have used prerequisite skills such as: function theory (forming $f(x + h)$ from the given $f(x)$); rationalizing the numerator of a rational expression (removing the square roots from the numerator of the expression (2.2) - necessary before we can apply limiting procedures) and algebraic simplification of the ensuing quantity, to arrive at the final form of the expression (2.3). As a calculus instructor, I am concerned with what happens after (2.3) i.e. evaluating the expression

$$f'(x) = \lim_{h \to 0} \frac{1}{\sqrt{x + h + 3} + \sqrt{x + 3}} \tag{2.4}$$

The process by which we arrive at (2.4) (i.e. the *precalculus*) is *assumed* to be familiar and well-known to each student. There just isn't enough time to undertake any significant review in this area. Knowing this before the course begins, will certainly save you time, effort and considerable stress, as well as allow you more time to concentrate on the important *new* material.

Step 2 *Once those all-important prerequisite skills have been identified, make sure they work!*
Once identified, it's important to make the necessary prerequisite skills *functional* i.e. make sure that you have these skills *at your fingertips* and ready to go. Don't be influenced by how you scored in previous mathematics courses - this can often lead to a false sense of security. Instead, make sure that you attain the required *fluency* in the necessary skills. Do this through practice. Perform a few sets of simple exercises in the relevant techniques. Perhaps five problems for each technique/skill. This will have two major benefits. Firstly, you will achieve the required fluency and secondly the mathematical side of your brain will re-awaken. These will be your *warm-up* exercises: simple but effective. You won't believe how much easier it will make the transition into your mathematics course.

Why do Highly Qualified Students Fail ?

Every year I like to address a class of new students about to enter a particular introductory mathematics course. It may be beginning calculus, linear algebra, geometry or something similar. I ask the students to raise their hands if they scored over 90%, 80% ..., etc, in the mathematics course *one level down* i.e. in the formal prerequisite course. Most of the students indicate scores of above 70% with a significant portion reporting scores of over 80%. Understandably, most of these students assume they are well-prepared to move onto the next level. I next ask the students to answer 10 multiple-choice questions in 20 minutes. These questions cover material from the corresponding essential prerequisite skills mentioned above. For example, the questions might deal with basic skills in factoring and simplification of algebraic expressions, solving inequalities, reading and using trigonometric formulas, quadratic equations, functions etc. After the 20 minutes has passed, I ask them to pass their answer sheet to the person sitting in front of them - and we grade them together. In all the years that I have conducted this 'experiment', I have never had more than 1% of the class score a perfect 10. The majority score around 5 or 6 correct answers out of 10 with an alarmingly significant number scoring below 5. Many simply run out of time while most have basically forgotten the relevant material. In either case, it is clear that the necessary prerequisite skills are not as fluent and effective as they are *required* or *assumed* to be.

At this point, as you can imagine, I have the students' full attention. Perhaps most significant is the fact that many students are not even aware that they are ill-prepared! High averages in prerequisite mathematics courses clearly often lull students into a false sense of security. In other cases, students have been absent from school for a significant period of time. Yet, because they have a formal *paper prerequisite*, they are admitted to a particular mathematics course but often do poorly because the necessary prerequisite skills are either absent or non-functional. Clearly, taking time *during* your mathematics course to reacquaint yourself with prerequisite techniques is wasteful, stressful and takes time away from the important new material. This is perhaps one of the main reasons why an excellent performance at one level does not necessarily translate into a similar performance at the subsequent level - and why seemingly highly qualified students do not perform to expectations.

Finally, I explain to the students exactly what has been discussed above - that a *paper prerequisite* is worth only the value of the paper if it doesn't *work*. We discuss the two-step procedure mentioned above and I suggest sets of practice problems/examples (depending on the particular mathematics course) to get their prerequisite skills back in working order.

Every year, many students return after their mathematics course to let me know how this particular intervention 'turned things around' for them. They also mention how, on applying these simple preparation techniques to other mathematics (and even non-mathematics related) courses, they have seen significant improvement in performance. This is perhaps the most rewarding aspect of my profession.

Chapter 3

HOW TO BE EFFECTIVE DURING YOUR MATHEMATICS COURSE

In this chapter, we demonstrate how to maximize your effectiveness in each of the various aspects of a typical mathematics course. We discuss efficient procedures for making the most of lectures and classroom time, how best to use your textbook, effective study habits leading to maximum performance in assignments/homework and how to get help - in particular, how to be most effective when getting help from your instructor.

3.1 MAKING THE MOST OF CLASS TIME

In general, there are three different class formats used in schools, colleges and universities:

Lectures
Lectures with Student Participation - Active-Learning
Tutorials, Seminars or Laboratories (Labs)

Lectures

Lectures are perhaps more common in colleges and universities than at the school-level. In a conventional lecture, there is very little interaction between the instructor and the students. Normally, the instructor will present the relevant information and the student will listen and take notes. The student's primary emphasis in a lecture, should be on *information-gathering*. When attending a lecture, one should bear in mind the following.

The Instructor has Prepared the Lecture with the Student's Objectives in Mind

Your instructor has most-likely spent several hours preparing the lecture material. For you to get the most out of a lecture, you should be aware of the instructor's objectives during the lecture. For a particular mathematics course, they include:

1. To extract the *relevant* information from the many textbooks on this subject and present it in the most clear, concise and understandable way possible.

2. To ensure that sufficient material has been covered to allow you to solve all the problems on any assignments/homeworks and any forthcoming examinations.

3. To present explanations and illustrative examples of relevant techniques (most often, understanding how to solve problems in mathematics begins with following and mimicing the steps used by your instructor).

4. To demonstrate effective procedures for solving problems and to indicate the standard required/expected in an assignment or examination.

5. To suggest relevant, targeted practice material.

6. To inform students of all other relevant information pertinent to the course and/or to the examinations.

In other words, lectures are a *primary source of* **relevant** *information.* It would be a mistake to try and learn-as-you-go in a lecture environment. The material often comes too fast to be 'digested' while it is being presented. Instead, accept that a lecture is an information-gathering session and concentrate on *effective note-taking.* Try to write down everything the instructor writes (following the same procedure and using the same notation) and take note of anything relevant he or she might say. Remember, the instructor makes up the course work and the examinations and the lecture material has been chosen specifically with these in mind. This means that the material presented in a lecture is targeted, relevant and extremely important - for all aspects of your course i.e. learning, assignments/homeworks and the all-important examinations.

Within 24 hours of the lecture (preferably as soon as possible - while the material is still fresh in your mind), you should re-read your notes, re-write any sections that are not clear and take note of anything you do not understand or any questions arising. Next, ask your instructor about these particular points/questions. At the end of this process you will have a clear and concise set of lecture notes - this will benefit you tremendously when returning to these notes when, for example, you review for a test or an assignment (there is nothing worse than trying to understand 'scratchings' taken at a lecture given several weeks previously - and the instructor is usually not terribly keen to return to material he thought he (and you!) had covered some time ago!).

Finally, lectures are usually not the best place to ask questions. There may be a few spare minutes at the end of a lecture but usually the instructor is concerned with getting to the next class or vacating the lecture hall for the next lecture. It is best for you and your instructor if you make a separate appointment to have your questions answered (see Section 3.4). The most effective students tend to *save* all their questions for one weekly appointment and use this information to fine-tune the whole week's lecture notes (the bulk of re-writing and understanding is performed as soon as possible after each lecture, as described above).

Lectures with Student Participation - Active-Learning

This is the format most often used in schools and institutions dealing with pre-college material. However, it can also be found in colleges and universities, particularly when class-size is relatively small. The material is covered using a combination of lecture, discussion, and question-and-answer sessions. Depending on the size of class, students may also be organized into groups of 2 - 5 and questions addressed either to the group or to the individual. This type of class has the student much more actively involved in his or her own learning (than say a lecture) and is usually much more effective than the typical lecture. The material is usually presented at a slower pace with time taken for reflection and discussion. Much of what has been said above (for lectures) applies here except for the fact that the decrease in the lecture component and the increase in the discussion and participation component of the class now give the student more time to understand the material as it is presented. Consequently, the student should be prepared to answer (and ask) questions. For example, in class, the instructor may work through a 'live' example, asking students questions such as:

- What's the next step ?

- Where have I gone wrong in the previous step ?

- How would I adjust this procedure to deal with a more general case ?

Tutorials, Seminars or Laboratories (Labs)

These are usually smaller discussion groups offered in conjunction with the usual series of lectures. The main objectives in a tutorial, seminar or laboratory (henceforth referred to as a tutorial) are

1. To reinforce lecture material with more worked examples, practice problems and discussion/active participation (remember, we learn mathematics by seeing examples and doing exercises).

2. To provide an opportunity for the student to ask questions and discuss points made in the lectures.

3. To discuss supplementary course materials. For example, to provide hands-on experience of mathematical software packages and their use in related areas.

11

Usually, the instructor will assign a particular set of problems to each tutorial. These problems are chosen to reinforce lecture material and to give practice for a forthcoming assignment/homework or test. Each problem set is almost always assigned well before the corresponding tutorial. The reason for doing so is to allow the student preparation time and consequently to use the tutorial to *ask questions*. You should think of a tutorial as *free instructor time* - an opportunity for you to ask as many questions as possible on specific examples from your mathematics course. As such, you must **prepare appropriately** i.e. *never* attend a tutorial without first having attempted each assigned problem, noted any difficulties and made a list of specific questions for your instructor. The most effective students will have thought through each problem *ahead of time* - so that they can spend as much time as possible *asking questions* in the tutorial. The least effective students show up unprepared, usually without having looked at the assigned material. As a result, they spend most of the tutorial time trying to figure out details which they could have considered in their own time (e.g. at home or in the library). They almost never ask questions (they are too busy doing what they should have done beforehand) and consequently waste the opportunity to ask questions. There is no doubt that these students will also arrive (eventually) at a list of questions related to the tutorial problems (the problems are designed to be challenging and thought-provoking) but they will then have to arrange (if possible) for a separate period of time to ask these questions. Never forget that an opportunity to ask course-related questions of someone who knows what you need to know, is an extremely valuable commodity - one which should never be wasted!

Those of us with experience of lawyers know that you never arrive for an appointment without knowing exactly what you need to know. The time spent in such a consultation is expensive and is directed towards discussing only relevant facts that cannot be discussed without your lawyer present. Consequently, things that you can do yourself, in your own time, you do *before* the appointment. This way, you pay less and maximize your effectiveness. Think of tutorial time accordingly. In the world of academics, access to an instructor is extremely valuable, particularly when *you* can choose which questions to ask.

3.2 THE COURSE TEXTBOOK

Most mathematics courses require that you buy a specified textbook. This textbook is chosen by your instructor to satisfy several requirements. They are:

- To provide a back-up source of course material. In this respect, the textbook will include the large majority (and more) of the material covered in your mathematics course.

- To provide a source of illustrative examples.

- To provide a good supply of practice problems.

It is important to understand that the textbook complements your mathematics course and is in no way a substitute for lectures and tutorials/seminars/labs. To understand why, consider a situation in which a student enrolled in (for example) a beginning calculus course is given the required textbook, told to study the first five chapters and then show up to class only for class tests and examinations. This student would no doubt encounter the following difficulties:

- The first five chapters of the textbook contain a wealth of information - the student can't hope to learn all of it. Which is most important? Which results can the student omit without worrying about whether or not they will appear on the examinations? What are the main results of each section? Each chapter?

- There are numerous practice exercises at the end of each section. The student does not have the time to work through every problem. Which of the problems are most important (for e.g. examination purposes)? Which of the problems are not relevant for this particular presentation of calculus (i.e. often, beginning calculus is taught differently depending on whether you are in business, science or engineering)? Which of the problems are of examination standard? Which of the problems are similar to the type likely to be asked on a test?

Without the instructor's guidance, it is extremely difficult to target relevant material from the textbook. One of the main reasons for attending lectures is that the instructor has spent considerable time and effort *targeting relevant* material, examples and practice problems. In lectures and tutorials, the instructor will tell you what's important and what isn't; what is examinable material and what isn't; which examples best illustrate the theory and which don't and which practice problems will give you the most effective preparation for assignments, class tests and examinations.

This is not to say that the text book is not important. It is an integral (no pun intended!) part of any mathematics course. However, there are ways to make sure that the textbook works *for* you in conjunction with the usual series of lectures and tutorials.

How To Make Effective Use of the Textbook

- Use the lecture material as your guide to what's relevant. If you need a 'second opinion' (alternative explanation) or more detail on a particular result, consult the text. The alternative (or more detailed) explanation may be clearer to you.

- Again, using the class material as your guide, use the text as a source of more illustrative examples of the theory covered in class. The more examples you see, the more likely it is that you will understand the basic concepts - and that you will be able to apply the procedure to a *new* problem.

- Use the text as a supply of additional practice problems - but target problems 'similar' to the ones solved in class or appearing on assignments/homeworks or on tests and examinations. This gives you *smart practice* i.e. relevant practice in the areas deemed to be important by your instructor. In this respect, it is always a good idea to ask your instructor to point out which problems in the textbook are most relevant.

3.3 ASSIGNMENTS/HOMEWORK

Why are assignments important? Certainly, one reason is that they can be worth anything from 10% to 20% of the final course grade. Another, perhaps more important, reason is illustrated in Figure 3.1. The data in Figure 3.1 represents average (final) examination scores for groups of students missing up to a maximum of 5 assignments in a particular mathematics course. It is interesting to note that the average score among those students submitting all 10 assignments was close to 70% while that for those missing even 4 assignments was less than 50% !

It is clear that missing assignments also affects examination performance! To see why, we need only recall the basic principle that *we learn mathematics by seeing examples and doing relevant exercises.* Assignments are vehicles for relevant and targeted practice! In other words,

1. Assignments tell you *where* you should concentrate your practice i.e. they identify relevant practice problems in relevant areas - as suggested by your instructor!

2. Assignments identify problems which best illustrate key concepts to enhance understanding.

3. Assignments identify problems similar to those which may be asked on examinations- the instructor is already giving clues!

4. Assignments identify required standards and expectations i.e. the problems chosen indicate the standard expected in a particular mathematics course.

5. In mathematics, working through relevant practice problems allows you to see patterns and note repetitions in solution techniques making subsequent problems easier to solve. Assignments provide an excellent source of such problems.

Figure 3.1: Significance of Assignments

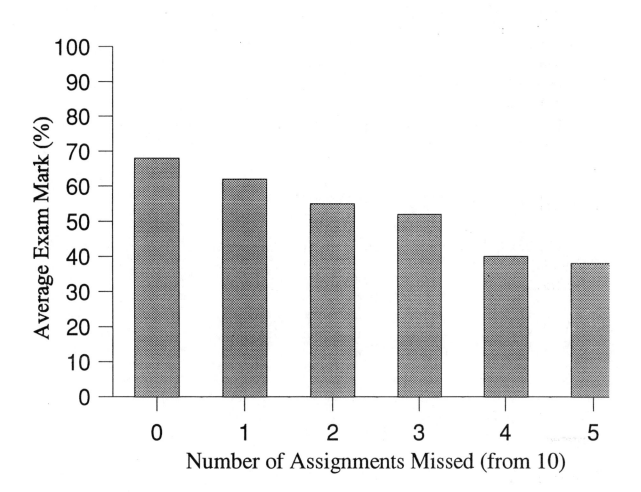

How to be Effective When Writing Assignments

Think of assignments as having three main purposes:

- To get that 10 - 20% of the course grade allocated to assignments.

- To get *practice* in relevant techniques

- To help you review for examinations.

All three of these objectives depend on how you present solutions to assignment problems i.e. what you write and how it's written. We demonstrate using the following examples. Again, as in Example 1.1, the subject matter is not important here - only the procedure and the presentation. Whatever is said with respect to the following examples is true for any problem in any area of mathematics. Example 3.3.1 is chosen from a beginning calculus course, Example 3.3.2 from trigonometry and Example 3.3.3 from basic algebra.

Example 3.3.1. Find the equation of the tangent to the curve represented by

$$y = f(x) = x \sin x$$

at the point $(\frac{\pi}{2}, \frac{\pi}{2})$.

Solution. The equation of a line with slope m, passing through the point (a, b) is given by

$$y - b = m(x - a) \qquad (3.3.1)$$

(see Calculus Book, page 332). Here, $(a, b) = (\frac{\pi}{2}, \frac{\pi}{2})$ (given in the question). Regarding the slope m,

$$m = f'(\frac{\pi}{2})$$

That is, the slope of the tangent to the curve $y = f(x)$ at the point $(\frac{\pi}{2}, \frac{\pi}{2})$ can be found by differentiating the function $f(x)$ and evaluating the resulting derivative at the value $x = \frac{\pi}{2}$ (see Calculus Book page 421). Using the product rule for differentiating functions (Calculus Book page 453)

$$f'(x) = 1 \cdot \sin x + x \cos x$$

Hence,

$$
\begin{aligned}
m &= f'(\frac{\pi}{2}) = 1 \cdot \sin\frac{\pi}{2} + \frac{\pi}{2}\cos\frac{\pi}{2} \\
&= 1 \cdot 1 + \frac{\pi}{2}(0) \\
&= 1
\end{aligned}
$$

Finally, from (3.3.1),

$$
\begin{aligned}
y - \frac{\pi}{2} &= (1)\left(x - \frac{\pi}{2}\right) \\
y &= x \qquad\qquad ans
\end{aligned}
$$

Example 3.3.2. Show that, for appropriate values of x,

$$
\frac{1 - \cos^2 x}{\cos^2 x \tan x} = \tan x \tag{3.3.2}
$$

Solution. We use the following trigonometric identities (see Trig. Book page 52) to reduce the left-hand side to the right-hand side.

$$
\begin{aligned}
1 - \cos^2 x &= \sin^2 x \tag{3.3.3} \\
\tan x &= \frac{\sin x}{\cos x}
\end{aligned}
$$

Consider then, the left-hand side of (3.3.2).

$$
\begin{aligned}
\frac{1 - \cos^2 x}{\cos^2 x \tan x} &= \frac{\sin^2 x}{\cos^2 x \tan x} \qquad \text{(using (3.3.3)} \\
&= \left(\frac{\sin^2 x}{\cos^2 x}\right)\frac{1}{\tan x} \\
&= \left(\frac{\sin x}{\cos x}\right)^2 \frac{1}{\tan x} \\
&= (\tan x)^2 \frac{1}{\tan x} \qquad \text{(using (3.3.3))} \\
&= \tan x \\
&= \text{right-hand side}
\end{aligned}
$$

Example 3.3.3. Solve the following equation

$$
x^2 - 4x + 2 = 0 \tag{3.3.4}
$$

17

Solution. The equation (3.3.4) is a quadratic equation. In general, the quadratic equation

$$ax^2 + bx + c = 0, \qquad a, b, c \text{ constants}$$

has two solutions given by (see Algebra Book page 23),

$$x = \frac{-b \pm \sqrt{b^2 - 4ac}}{2a} \tag{3.3.5}$$

Noting that, in (3.3.4), $a = 1$, $b = -4$ and $c = 2$, we obtain, from (3.3.5),

$$
\begin{aligned}
x &= \frac{-(-4) \pm \sqrt{(-4)^2 - 4(1)(2)}}{2(1)} \\
&= \frac{4 \pm \sqrt{16 - 8}}{2} \\
&= \frac{4 \pm \sqrt{8}}{2} \\
&= \frac{4 \pm 2\sqrt{2}}{2} \qquad (\text{since } \sqrt{8} = \sqrt{4 \cdot 2} = \sqrt{4} \cdot \sqrt{2} = 2\sqrt{2}) \\
&= 2 + \sqrt{2} \quad \text{or} \quad 2 - \sqrt{2}
\end{aligned}
$$

In other words, (3.3.4) has solutions

$$x = 2 + \sqrt{2} \qquad \text{and} \qquad x = 2 - \sqrt{2}$$

In each of Examples 3.3.1 - 3.3.3, we have followed a clear, logical procedure in arriving at the correct answer. In addition, we have recorded all the steps in the calculation and the precise reasoning used to arrive at each step. Note that we have also a record of exactly where each piece of subsidiary information can be found (in the textbook or otherwise).

As far as the first purpose of an assignment is concerned (to get *practice* in relevant techniques), the procedure by which you arrive at the correct answer is just as important as the answer itself. It is exactly the development and repeated application of a set *procedure* which will allow you to begin to see patterns and repetitions in solutions - which is the main reason for *practice* in mathematics. For example, if you were now asked to find all solutions of the equation

$$2x^2 - 5x + 1 = 0$$

you could mimic the solution to Example 3.3.3. The reasoning is clear and easy-to-follow because of the effort invested *first time around.* In this respect, assignment solutions should resemble *maps* or *recipes*, recording where and how to proceed when you want to repeat the same or similar calculation.

The following are additional reasons to record as much as possible of your solution procedure (bear in mind the other two purposes of an assignment - to get that 10 - 20% of the course grade allocated to assignments and to help you review for examinations):

- A good step-by-step solution procedure enhances understanding - *it's like teaching/explaining the material to yourself.*

- It may take some 'hard thinking' to solve certain problems - recording the steps will save time next time you encounter a similar exercise, leaving you more time to practice with new problems (the *map* concept).

- The person grading the assignment will be impressed by your clear and logical thinking (indicating understanding) - and will find it much easier to allocate part-marks when, for example, an arithmetic slip has led to an incorrect final answer.

- When it comes to examinations (remember, assignment questions are excellent as practice examination problems), a detailed approach to assignment solutions allows for *reviewing* as opposed to *relearning* i.e. there's nothing worse than trying to remember how you did something and then having to spend the same time (again) relearning how to do it - just keep a record! - consult your *map* or *recipe*.

- Detailed solution procedures will help to organize your thoughts and develop effective problem-solving techniques.

There is no doubt that more detailed solutions will, at first, take more time. This extra time should be thought of as an investment in the future. Soon, the 'step-by-step' approach will become automatic and you will have developed an effective approach for solving many different kinds of problems.

Using Posted Solutions to Assignments

In many cases, an instructor will provide detailed solutions to assignment problems. It is always a good idea to consult these solutions, irrespective of your grade on a particular assignment. There are two basic reasons for doing so.

1. By consulting the instructor's solutions, you see exactly what constitutes a proper/correct solution (procedure) in the eyes of the instructor (usually the person who grades the *examinations*). This way, you know exactly what's expected.

2. In some cases students achieve perfect scores on assignments despite making errors and often not using the most effective solution procedures. In my experience, this is almost always attributed to the fact that teaching assistants grading

assignments may not always examine every detail of each solution. Remember, an assignment is not just a way of accumulating marks (recall the three purposes of assignments, mentioned above), it is also a way of developing efficient solution procedures and an important component of exam-preparation. For these reasons, you should always check your assignment against the instructor's solutions - just to make sure that you are doing what you should be doing. When it comes to course examinations, it is almost always the instructor who does the grading. I cannot tell you the number of times I have been approached by students (after I have graded their examination) and asked why marks had been deducted from a particular question when they "did it the same way in the assignment and got full marks !". I answer by reminding them why I post solutions to assignments!

Doing Assignments Yourself Versus in a Group

Students working in groups enjoy many advantages including:

- Seeing and learning alternative problem-solving strategies.

- Learning from other students - adds to and improves the overall learning experience.

- Individual students who get stuck may give up but groups tend to keep the momentum going.

- Active/cooperative learning i.e. people in groups tend to verbalize their thinking which adds to the problem-solving process.

- Improvement in self-esteem and a decrease in anxiety levels (talking about things relieves anxiety).

However, when all is said and done (unless you are required to submit a group assignment) the final submitted assignment must reflect your *own* understanding - you will not have the benefit of a group in the examinations! For this reason it is perhaps best to adopt a 'middle-of-the-road' approach. By all means formulate and discuss ideas and opinions in collaboration with your colleagues. Think of this as part of the *information-gathering* process. Use this information to put together your *own* assignment. Do not submit anything you do not understand - even if it is correct. First, find out *why* and *how* and then (with the three objectives of an assignment in mind) write and submit your assignment.

3.4 GETTING HELP FROM YOUR INSTRUCTOR

It is interesting that many students (particularly new students) will not ask for help from their instructor for fear that they are 'bothering' him or her. What these students fail to recognize is that the instructor fully expects students to come to his/her office to ask questions - it is one of his/her professional duties. Having said that, however, there are certain ways to ensure that the time spent in consultation with your instructor is used as effectively as possible.

Firstly, it is important to recognize that an instructor has many duties to perform. Consequently, he or she is not always available to answer questions. Most instructors will assign *Office Hours* to a particular course. This is a block of time set aside for the sole purpose of answering students' questions. During this time, the instructor is happy to discuss any aspect of the lecture material, the assignments or any other questions relating to the course. If, for some reason, you cannot meet with the instructor during the assigned office hours, you can ask for a separate appointment at a mutually convenient time.

Whichever way you choose to meet with your instructor, there are certain guidelines you should follow - to ensure the most effective use of time spent in consultation. The follow are the 'Do's and Don'ts' of getting help from your instructor.

DO:

- Arrive *on time* i.e. either during the assigned office hours (preferably not during the last minute) or at the time of your pre-arranged appointment.

- Arrive prepared and organized i.e. know *exactly* what you need to ask. Prepare a list of questions beforehand (if necessary) and bring any necessary supporting materials e.g. textbook, list of questions etc.

- If you intend to ask a question on a particular problem, bring your work *so far* i.e. have a written account (clear and methodical as in Examples 3.3.1 - 3.3.3, above) of *your attempt* at the problem indicating where you 'got stuck' and what method/argument you have used to arrive at this point.

- Keep your questions targeted and focused.

- Be professional, polite and courteous at all times. If you do not understand a particular explanation, ask (politely) to have it repeated.

- Ask or encourage the instructor to *write down* any help or explanation - that way you have a record of your visit and, later, an opportunity to review what was said in a more relaxed atmosphere. In this respect, it's always a good idea to have a pen and a blank piece of paper available - to present to the instructor when he/she begins to explain something. Mathematics is primarily a written language - we remember very little of 'spoken explanations' - so make it easy for the instructor to 'write it down'.

DON'T:

- Show up outside office hours or pre-arranged appointments. More often than not you will have a wasted journey. Instructors allocate their time carefully according to their various duties.

- Ask unfocused, non-specific questions such as "How do I do Problem 3 of the assignment ?". This gives the impression that you have not thought about the problem and wish the instructor to do your work for you! Instead, present your (clear and logical) attempt at the question and ask your instructor's advice on, for example, the next step or where you may have gone wrong. The instructor is more likely to provide constructive assistance in this scenario.

- Ask for help on a problem which you have not thought about - again, this is basically asking the instructor to solve the problem for you. First try the problem yourself !

- Argue with the instructor. Maintain a good working relationship.

- Be afraid to ask as many questions as you need to ask. Just remember to be polite and professional and follow the points made above. More often than not, instructors will interpret activity during office hours etc as being indicative of a student's conscientious effort to do well in the course. This is usually to the student's advantage when the instructor comes to assessing overall performance in the course.

Following these simple rules will ensure that you get your questions answered quickly and effectively and that you maintain a good working relationship with your instructor.

3.5 OTHER SOURCES OF HELP

Someone once said that the three golden rules for success in mathematics are "*Ask! Ask! Ask!*" There is no doubt that we will all encounter difficulties in mathematics so it makes sense to ask for help whenever it's made available. Usually, the following sources of help are available as part of your mathematics course:

1. Your instructor (see Section 3.4 above).

2. Help Sessions - In many cases teaching assistants will make themselves available to answer questions either in a separate office or in a specially designated room. The same principles discussed in Section 3.4. apply here. These sessions are often extremely busy, consequently, it makes sense to go prepared with questions in hand.

Other sources of help, such as tutors and study guides usually carry an extra fee. They will be discussed later in Chapter 6.

Finally, it is worth noting that student advisors and guidance counsellors will offer non-technical advice on various different aspects of the academic experience. These people are very experienced in the type of problems you are likely to face as a new or continuing student and are usually excellent sources of information.

Chapter 4

PROCEDURES FOR EFFECTIVE PROBLEM-SOLVING

We have already noted that examples and exercises are essential components of maximizing performance in any particular mathematics course. Consequently, a significant amount of course-time (lectures, tutorials/labs and assignments) is devoted to worked examples and relevant practice problems. In Chapters 1 and 3 we emphasized the importance of *presentation* in developing effective problem-solving techniques i.e. how an effective solution requires that you demonstrate a clear, logical and organized procedure. In this chapter, we examine actual problem-solving strategies i.e. effective procedures for solving mathematics problems. We illustrate our ideas with worked examples from different areas of mathematics.

Basically, two types of problems are encountered in mathematics courses:

TYPE A *Those which require mainly application of known techniques but minimal thinking.* In other words, problems which require you to repeat from memory, know the meaning of certain key concepts and apply established course material to new situations. These are also known as 'plug-and-chug' problems - solved by applying a formula and set procedure.

TYPE B *Those which require mainly thinking but minimal application of established techniques.* These problems exercise the higher-level thinking skills and, as such, are often more difficult than TYPE A problems. They usually involve some *mathematical modelling* followed by *evaluation* and *application* of selected mathematical techniques, usually from course material.

The large majority of problems in most introductory mathematics courses (in fact, up to senior undergraduate level) are of TYPE A. TYPE B problems, although most common in graduate-level courses, appear also in most high-school, college and university introductory mathematics courses - usually in the form of *word problems*. In the following sections, we examine procedures used to solve each type of problem.

4.1 SOLVING PROBLEMS WHICH REQUIRE MAINLY APPLICATION - TYPE A

Problems of this type serve mainly to reinforce lecture material. As such, they are chosen to encourage *practice* in the relevant techniques - to actually *do the thing*. In mathematics, understanding the material is only one part of the learning process. The other (more important) part is concerned with actually applying the material by *yourself* - in practice problems. This latter part is the one which is most often neglected when learning mathematics. There are always people that will read through a well-written solution to a problem (from, for example, the textbook, class notes or another student's assignment), understand the solution and believe that they then know how to 'do it'. Unfortunately, this is almost always the wrong conclusion - they *think* they know how to do it but in actual fact, they don't! To use some simple analogies, we learn to drive *on the job* i.e. by actually driving! We learn to swim by getting into the water and actually practicing or performing the movements! *Watching* someone else drive or swim and *understanding* what they are doing is not sufficient to allow us to perform those same movements, in the same way, by ourselves - we need to practice first! In mathematics, we are required to *perform* i.e. to apply the theory ourselves. Solving a problem 'from scratch', starting from a blank sheet of paper, is completely different from reading through the solution to the same problem.

Something else to remember about problems of TYPE A is that, since they are chosen primarily to practice a particular technique, by their very nature, they belong to distinct classes (each class associated with the technique being practiced) each of which identifies a particular solution procedure. Hence, the first step in an effective solution procedure involves identifying the class to which a particular problem belongs. For example, all problems involving finding the solutions of a quadratic equation follow a set procedure (as in Example 3.3.3) which makes use of the formula (3.3.5). Similarly, problems involving differentiation of products of functions follow a procedure which makes use of the product rule for differentiation (see Example 3.3.1) and so on.

The following is a systematic problem-solving procedure which is extremely effective for solving TYPE A problems (Recall that the large majority of problems you encounter in your mathematics course are of TYPE A). To begin with, imagine you have an assignment or practice problem in front of you.

1. Read through the problem carefully and classify the problem by identifying the (general) corresponding area discussed in class notes or textbook. For example, does the question deal with solving algebraic equations or functions or Pythagoras' theorem or perhaps differentiation or maybe trigonometric identities. Whichever it maybe, there will be a set procedure for solving the problem. Your job is to *find* and *mimic* that procedure. In classifying a problem which deals with more than one area, classify according to *primary* area. For example, a problem dealing with the equation of a tangent line to a curve represented by a given function, may be chosen primarily for practice in the techniques of differentiation even though some analytic geometry is used in arriving at the solution (see Example 3.3.1).

2. Once the general area has been identified, the classification has to be more specific. Read through the problem again and decide what (exactly) you are required to find. For example, if you are dealing with a (calculus) problem in

differentiation, are you being asked to find the derivative of a product, quotient or composition of functions? If the problem is concerned with matrices, are you finding the inverse or applying matrix multiplication procedures? If the problem deals with functions, are you being asked to find the domain and range or perhaps to sketch the corresponding graph? Get to the heart of the problem and identify the specific area/technique/formula/rule which will form the basis of the solution..

3. Once you know what you are looking for, go to your major source of information. This may be your class notes, textbook or whatever is being used as your main source of material. Look for the corresponding area and, more specifically, that dealing with the main formula/rule/technique that must be applied. Next, find a worked example *similar* to the one you are trying to solve, using that same formula/rule/technique . This is an extremely important part of the problem-solving process - there is no need for you to 're-invent the wheel' - all you need to do is adapt a successful solution procedure to your particular example. That's the nice thing about TYPE A problems - most have already been solved and those that haven't are similar to those that have (remember the main reason for solving TYPE A problems)!

4. Apply the known procedure to your specific example, following each step carefully. Remember to be clear and logical and to explain (to yourself) at each step exactly what you have done (see Section 3.3 to remind yourself of the reasons why this is important).

Now you have the frame and most of the body of the solution and your attention is focused on a specific area. If the problem is 'straightforward', you can now proceed to finish the solution. Occasionally, you will encounter difficulties ('get stuck') perhaps because the problem you are solving has a peculiar characteristic which makes it slightly different from the example you are following. This is OK - in fact, it's an important part of the learning process. If you find yourself in this situation, *do not give up*! Try to get around the problem by examining in a bit more detail the theory surrounding the example from the class notes or textbook. You will find that the harder you try, the deeper your understanding of the problem becomes! If, after spending some time thinking about things, you still cannot make any progress, get some help (see Sections 3.4 and 3.5). Often, all it takes is a little nudge in the right direction. In the large majority of cases in which I 'help' students with problem-solving, I spend no more than a few minutes with each student - usually we overcome the difficulty on the first examination of the student's (partial) solution. In fact, most of the time, the student, still actively thinking about the problem, will come up with the answer during the discussion. That's why it's important to use *help* as a *part* of the problem-solving process and not as the problem-solving process itself! Seeking help prematurely is extremely counter-productive!

In the following examples, we illustrate the above procedure using problems chosen from a range of mathematics courses.

Example 4.1.1. Simplify the following expression.

$$\frac{2x}{3x+2} - \frac{4x}{3x+1}$$

Preliminary thoughts:

1. General area: Simplification of algebraic expressions

2. More specifically: Rational expressions, taking common denominator and possibly some factoring.

3. Find a similar example in textbook or class notes.

4. Mimic that example to produce the correct solution as follows.

Solution.

$$\begin{aligned}
\frac{2x}{3x+2} - \frac{4x}{3x+1} &= \frac{2x(3x+1) - 4x(3x+2)}{(3x+2)(3x+1)} \\
&= \frac{6x^2 + 2x - 12x^2 - 8x}{(3x+2)(3x+1)} \\
&= \frac{-6x^2 - 6x}{(3x+2)(3x+1)} \\
&= \frac{-6(x^2 + x)}{(3x+2)(3x+1)} \\
&= \frac{-6x(x+1)}{(3x+2)(3x+1)}
\end{aligned}$$

The solution ends here since we have 'combined' both terms into a single term by taking a common denominator and we have factored the numerator and denominator of the resulting rational expression as far as possible.

Example 4.1.2. Find the domain and range of the following function.

$$f(x) = \sqrt{3 - x}$$

Preliminary thoughts:

1. General area: Function theory.

2. More specifically: Domain and range - function involves a square root.

3. Find a similar example in textbook or class notes.

4. Mimic that example to produce the correct solution as follows.

Solution.

Domain: All real values of x for which $f(x) = \sqrt{3-x}$ 'makes sense' (in real numbers, we cannot take the square root of a negative number). This requires that

$$
\begin{aligned}
3 - x &\geq 0 \\
3 &\geq x \\
x &\leq 3
\end{aligned}
$$

Finally, domain $= \{x \in \Re : x \leq 3\}$.

Range: If $x = 3$, $f(x) = \sqrt{3-x} = 0$. If $x < 3$, $f(x) > 0$ (Note that x cannot exceed 3). Consequently, the range is $[0, \infty)$.

Example 4.1.3. Solve using elimination

$$
\begin{aligned}
y + x &= 10 \\
2y - x &= 5
\end{aligned}
$$

Preliminary thoughts:

1. General area: System of linear algebraic equations.

2. More specifically: Two equations in two unknowns - use elimination.

3. Find a similar example in textbook or class notes.

4. Mimic that example to produce the correct solution as follows.

Solution.

$$
\begin{aligned}
y + x &= 10 \\
2y - x &= 5
\end{aligned}
$$

Add the two equations

$$
\begin{aligned}
3y &= 15 \\
y &= 5
\end{aligned}
$$

Substitute for y in any of the original two equations

$$
\begin{aligned}
y + x &= 10 \\
5 + x &= 10 \\
x &= 5
\end{aligned}
$$

Solution is $x = 5$, $y = 5$.

Example 4.1.4. Differentiate the following function of x.

$$
f(x) = \frac{\sin x \cos x}{x + 1}
$$

Preliminary thoughts:

1. General area: Differentiation of functions

2. More specifically: A quotient of functions but numerator involves a product of functions. Expect to use both quotient and product rules for differentiation and possibly trig. identities and factoring techniques.

3. Find a similar example in textbook or class notes - one that demonstrates the quotient rule.

4. Mimic that example to produce the correct solution as follows.

Solution.

$$
f(x) = \frac{\sin x \tan x}{x + 1}
$$

First use quotient rule

$$
\begin{aligned}
f'(x) &= \frac{(x+1)\frac{d}{dx}(\sin x \tan x) - \sin x \tan x \frac{d}{dx}(x+1)}{(x+1)^2} \\
&= \frac{(x+1)\frac{d}{dx}(\sin x \tan x) - (\sin x \tan x)(1)}{(x+1)^2}
\end{aligned}
\tag{4.1.1}
$$

To obtain $\dfrac{d}{dx}(\sin x \tan x)$, use the product rule for differentiation.

$$
\begin{aligned}
\frac{d}{dx}(\sin x \tan x) &= \sin x \frac{d}{dx}(\tan x) + \tan x \frac{d}{dx}(\sin x) \\
&= (\sin x)\left(\sec^2 x\right) + \tan x \cos x \\
&= (\sin x)\left(\sec^2 x\right) + \sin x \\
&= \sin x \left(\sec^2 x + 1\right)
\end{aligned}
$$

29

Returning to (4.1.1),

$$f'(x) = \frac{(x+1)\sin x\,(\sec^2 x + 1) - (\sin x \tan x)(1)}{(x+1)^2}$$

$$= \frac{(x+1)\sin x\,(\sec^2 x + 1) - \sin x \tan x}{(x+1)^2}$$

$$= \frac{\sin x\,[(x+1)\,(\sec^2 x + 1) - \tan x]}{(x+1)^2}$$

Example 4.1.5. Find

$$\int \frac{x}{\sqrt{x^2+1}}dx$$

Preliminary thoughts:

1. General area: Integration or Anti-differentiation of functions

2. More specifically: Integration of a function in the form of a quotient. The derivative of *part* of the denominator appears on the numerator - Use the substitution rule for integration - and perhaps some factoring or simplification.

3. Find a similar example in textbook or class notes - one that demonstrates the substitution rule.

4. Mimic that example to produce the correct solution as follows.

Solution.

$$\int \frac{x}{\sqrt{x^2+1}}dx$$

Let $u = x^2 + 1$. Then, $du = 2x\,dx$. Substitute into the integral.

$$\int \frac{x}{\sqrt{x^2+1}}dx = \frac{1}{2}\int \frac{du}{\sqrt{u}}$$

$$= \frac{1}{2}\int u^{-\frac{1}{2}}du$$

$$= \frac{1}{2}\left[\frac{u^{\frac{1}{2}}}{\frac{1}{2}}\right] + C$$

$$= u^{\frac{1}{2}} + C$$

Here, C is an arbitrary constant of integration. Finally, substitute back for u using $u = x^2 + 1$ to obtain

$$\int \frac{x}{\sqrt{x^2+1}}dx = (x^2 + 1)^{\frac{1}{2}} + C$$

Examples 4.1.1 - 4.1.5 illustrate how the above problem-solving procedure is independent of any particular area of mathematics - it applies equally well to problems in all areas of mathematics.

4.2 SOLVING WORD PROBLEMS - TYPE B

As the name suggests, *word problems* are (mathematics) problems posed in English rather than mathematics. Most real-world problems are first posed as word problems since they arise naturally from everyday descriptions of physical situations.

Word problems are generally regarded as more difficult than TYPE A problems, the reason being that before any mathematics can be applied to the solution of a word problem, the problem itself must first be translated from English into mathematics. This process is called *mathematical modelling* and is, by far, the most challenging part of the solution of a word problem. There are three stages involved in solving a word problem:

Stage 1 Mathematical Modelling.
Stage 2 Mathematical Analysis.
Stage 3 Interpretation of the (mathematical) solution in terms of the underlying physical problem.

In most introductory mathematics courses, Stage 1 is concerned with changing a word problem into one of TYPE A (i.e. translating the English into mathematics) which is then solved using the application of course material (Stage 2). Stage 3 recognizes that the (mathematical) solution has some physical significance (since word problems deal mainly with practical, *real-world* situations) and is concerned with interpreting the solution in the context of the original *real problem*. The following are typical examples of word problems.

Example 4.2.1. A mathematician, as a reward for performing some great service to the King, was given one wish. The mathematician answered with the following request (which most thought was very modest): she asked for one piece of gold to cover the first square of a chessboard, two to cover the second, four to cover the third, eight the fourth and so on until the entire chessboard was covered. How many pieces of gold would the king have to give the mathematician to cover the last square?

Example 4.2.2. The monthly profit P in dollars on the sale of x units of a certain toy is

$$P(x) = \frac{x^3}{3} - 3x^2$$

When is profit increasing and when is it decreasing?

Example 4.2.3. If the midpoints of the consecutive sides of *any* quadrilateral are connected by straight lines, prove that the resulting quadrilateral is a parallelogram.

Example 4.2.4. If 1800cm^2 of cardboard is available to make a box with a square base and an open top, find the dimensions of the box with largest possible volume.

Each of these examples *hides* a corresponding mathematics problem and the first step in solving any of them is to *reveal* that problem. By their very nature (mainly because of the mathematical modelling) word problems do not lend themselves to solution by *set procedure* i.e. there are no hard and fast rules which will guarantee success every time. However, there are a number of general steps which are extremely effective in Stage 1 of the problem-solving process.

- Read the problem carefully - at least twice.

- Try to identify

    (i) What you *want* i.e. the unknown.
    (ii) What you *have* i.e. what is *given*.

In this respect, look for *keywords*. This will allow you to 'cut through' the 'padding' and reveal the *significant* information.

- In many cases it might be useful to draw a diagram to help you see what is going on.

Next comes the difficult part i.e. the translation from English into mathematics. Despite the fact that word problems are not TYPE A problems, there *will* be some clues about the underlying mathematics to be used.

- Look for *keywords* related to mathematical concepts. For example:

Example 4.2.1 The words 'one', 'two', 'four' and 'eight' suggest a *sequence* of some sort.
Example 4.2.2 The words 'increasing' and 'decreasing' suggest *rates of change* and *derivatives.*
Example 4.2.3 The question is concerned with geometry so a diagram will be essential. The words 'parallelogram' and 'quadrilateral' suggest that we have to use some geometrical technique to solve this problem - perhaps *vectors.*
Example 4.2.4 The word 'largest' suggests some sort of *maximum or minimum value problem* which again will involve *differentiation.*

- Introduce a suitable mathematical notation - preferably similar to that used in class or in the textbook. Choose symbols for the unknown quantities and the given information e.g. V for volume, t for time etc.

- Use mathematics to relate **what you want to what you know** - a relation which will connect the given information and the unknown *and* enable you to calculate the unknown. This will be your mathematical model.

At the end of this final step, you should have a TYPE A problem. Next comes Stage 2 of the process. This time, however, you should be on familiar territory - just follow the procedure for solving TYPE A problems outlined in Section 4.1 above. Stage 3 of the process will involve some sort of physical interpretation of your (mathematical) result. In this respect, you should know the meaning of the various mathematics used in arriving at the final solution. This will allow you to rule out any (mathematical) solutions which do not make sense (physically) in the context of the situation described in the problem. This information is usually available from class notes, the textbook or the problem itself.

There is one significant factor which makes word problem solving 'easier' and more effective: **experience.** That is, the more word problems you attempt and solve, the more familiar you will become with the process of mathematical modelling, the identification of keywords and the way in which you link *what you want to what you know* in some relation which will enable you to find the unknown. These are all key ingredients of an effective (word) problem-solving process. Consequently, once again, it comes down to *practice* leading to pattern recognition, fluency and finally effectiveness in problem-solving.

Some Other Useful Ideas for Solving Word Problems

- **Similar Problems**

As in the case of TYPE A problems, it is often possible to find similar, related or analogous (solved) word problems in either class notes or the textbook. For example, these problems may have the *same keywords* as the problem you are trying to solve. In this case, use the solution to furnish clues on how to proceed when solving your particular word problem (particularly for the mathematical modelling).

- **Subproblems**

Sometimes it is necessary to split a more complicated problem into a series of simpler *subproblems* each with its own *subgoal.* Each subproblem is solved in turn and the complete solution is built-up from each of these (sub)solutions.

Mathematics is all about reducing complicated problems to simpler ones. Word problems are no exception!

- **Write it all Down!**

With word problems, it is even more crucial to record, on paper, each stage of the problem-solving process (see Section 3.3). Whenever you figure out how to (mathematically) model a physical situation, you should record exactly the reasoning that led to the model. This is usually the most difficult part of the solution and you may wish to apply similar reasonings to other future word problems (The great mathematician Descartes once said that "Every problem that I solved became a rule which served afterwards to solve other problems".). Also, a clear logical procedure will allow you to check each stage of the process if an error is found somewhere in the solution.

- **Identify the Section of the Course Material with which the Problem is Associated**

When all else fails, clues on the techniques to be used to model and solve a word problem can be found in some very obvious places. For example, if the problem comes from a textbook, ask yourself the following question. With which section of the textbook is this (or a similar) problem associated ? That is, in which particular section does this problem appear? For example, if the problem appears in a section dealing with differentiation, you can be confident that the problem will be solved using ideas from differentiation. If the problem comes not from the textbook but from class materials (e.g. a handout), note which class material it is intended to reinforce and which particular topic it is intended to cover. These all provide clues to the mathematics which should be used to solve word problems.

- **Ask for Help**

If you find yourself 'completely stumped' and unable to start a particular word problem, it is best to ask for help from someone with experience in solving this type of problem e.g. your instructor (see Section 3.4) or otherwise (see Section 3.5). Often, identifying the keywords or just a simple clue on 'where to look' is sufficient to *get the ball rolling*.

- **Don't be Fooled by Solutions**

Perhaps more than any other type of problem-solving procedure, an effective procedure for solving word problems requires actual *hands-on* experience. It is crucial that you attempt and solve word problems *yourself*. It would be a mistake to read through someone else's solution and conclude that you now know 'how to do it'. The solutions will *always* look easier than expected. This is due primarily to the fact that the most demanding (and significant) component of the solution of word problems is the mathematical modelling. This usually requires considerable effort in thought and concentration, the majority of which is not reflected in the final solution.

In what follows, we illustrate the above problem-solving principles by discussing each of Examples 4.2.1 - 4.2.4.

Example 4.2.1. A mathematician, as a reward for performing some great service to the King, was given one wish. The mathematician answered with the following request (which most thought was very modest): she asked for one piece of gold to cover the first square of a chessboard, two to cover the second, four to cover the third, eight the fourth and so on until the entire chessboard was covered. How many pieces of gold would the king have to give the mathematician to cover the last square?

Solution. Cut through the irrelevant information (the 'padding') and get to the heart of the problem.

We *want*: the number of gold pieces required to cover the final square in a chessboard.

We *know*: that one piece covers the first, two the second, four the third, eight the fourth and so on.

A diagram will help us to note a *pattern*:

Square #	1	2	3	4	...	...
# of pieces of gold	1	2	4	8	...	...

Connect what we have to what we want, using some mathematics.

The number of pieces of gold placed on a particular square follows a *sequence* according to the above table. After some consideration, we note that the number of pieces on subsequent squares increases with powers of the number 2. That is,

Square #	1	2	3	4	....	n
# of pieces of gold	2^0	2^1	2^2	2^3	....	?

35

Let a typical square on the chessboard be labelled the nth square. For the nth square then, we might guess that (noting the pattern from the table) the number of gold pieces required on the nth square is given by 2^{n-1} (notice from the table that the power of 2 is always one less than the Square #). In the language of sequences, we have a geometric sequence with common ratio 2.

If we denote the number of gold pieces on the nth square of the chessboard by u_n (the nth-term of the geometric sequence), we can write the following formula for u_n.

$$u_n = 2^{n-1}$$

This is our mathematical model. Noting that there are 64 squares on a chessboard (if you didn't know this, think of it as part of the research required by the problem!), the number of gold pieces required for the last square is given by

$$u_{64} = 2^{64-1} = 2^{63} \qquad \text{(more than a million, million, million!)}$$

Example 4.2.2. The monthly profit P in dollars on the sale of x units of a certain toy is

$$P(x) = \frac{x^3}{3} - 3x^2$$

When is profit increasing and when is it decreasing?

Solution. Consider first, what we *want*, what we *know* and try to *connect* the two with information that enables us to calculate the unknown.

We *want*: to know when profit P is increasing/decreasing.

We *know*: that the profit P is given by the quadratic expression $P(x) = \dfrac{x^3}{3} - 3x^2$.

To connect *what we want to what we know*, we note that a function increases/decreases when its first derivative is positive/negative. Find the first derivative of $P(x)$.

$$\begin{aligned}
\frac{dP}{dx} &= \frac{d}{dx}\left(\frac{x^3}{3} - 3x^2\right) \\
&= x^2 - 6x \\
&= x(x - 6)
\end{aligned}$$

This is the mathematical model which will enable us to find what we *want*. To find when $\dfrac{dP}{dx}$ increases/decreases, we solve the following inequalities.

$$\begin{aligned}
\frac{dP}{dx} &= x(x - 6) \leq 0 \\
\frac{dP}{dx} &= x(x - 6) > 0
\end{aligned}$$

We find that

$$\frac{dP}{dx} = x(x-6) \qquad \begin{cases} \leq 0, & 0 \leq x \leq 6, \\ > 0, & x < 0, \quad x > 6 \end{cases}$$

It follows that P increases when $x < 0$ and $x > 6$ and decreases when $0 < x < 6$. Before we translate these results into the information required by the problem, bearing in mind Stage 3 of the problem-solving process, we note that since x represents the number of units sold, x cannot be negative. Consequently, we conclude that profit is increasing whenever sales are greater than 6 units (we rule out the physically meaningless case of $x < 0$ - this is still a valuable *mathematical* solution but of no interest in the physical context of the word problem) and decreasing when sales are between 0 and 6 units.

Example 4.2.3. If the midpoints of the consecutive sides of *any* quadrilateral are connected by straight lines, prove that the resulting quadrilateral is a parallelogram.

Solution. This question is much more abstract than any of the other examples. To see what we want and what we know, it is best to begin by drawing a diagram.

Figure 4.1

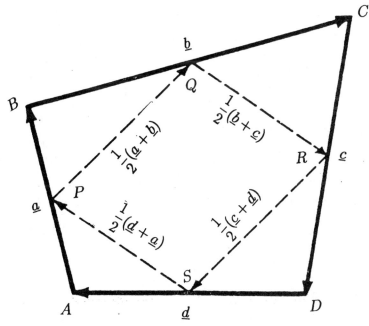

First set up the required notation. Let $ABCD$ be *any* given quadrilateral, P, Q, R and S, the midpoints of its sides and $\underline{a}, \underline{b}, \underline{c}$ and $\underline{d}$, the vectors $\overrightarrow{AB}, \overrightarrow{BC}, \overrightarrow{CD}$ and $\overrightarrow{DA}$, respectively (refer to Figure 4.1). We need to show that $PQRS$ forms a parallelogram i.e. that its opposite sides are parallel.

We *know* (see Figure 4.1):

$$\overrightarrow{PQ} = \frac{1}{2}(\underline{a} + \underline{b})$$

$$\overrightarrow{QR} = \frac{1}{2}(\underline{b} + \underline{c})$$

$$\overrightarrow{RS} = \frac{1}{2}(\underline{c} + \underline{d})$$

$$\overrightarrow{SP} = \frac{1}{2}(\underline{d} + \underline{a})$$

$$\underline{a} + \underline{b} + \underline{c} + \underline{d} = \underline{0} \qquad (4.2.1)$$

We *want* to show that:

$$\overrightarrow{PQ} \text{ is parallel to } \overrightarrow{SR}$$
and
$$\overrightarrow{QR} \text{ is parallel to } \overrightarrow{PS}$$

Using vector algebra, we can *connect* what we know to what we want, as follows:

$$\overrightarrow{PQ} = \frac{1}{2}(\underline{a} + \underline{b}) \underset{(4.2.1)}{=} -\frac{1}{2}(\underline{c} + \underline{d}) = -\overrightarrow{RS} = \overrightarrow{SR}$$

$$\overrightarrow{QR} = \frac{1}{2}(\underline{b} + \underline{c}) \underset{(4.2.1)}{=} -\frac{1}{2}(\underline{d} + \underline{a}) = -\overrightarrow{SP} = \overrightarrow{PS}$$

It follows that, in fact, $\overrightarrow{PQ} = \overrightarrow{SR}$ and $\overrightarrow{QR} = \overrightarrow{PS}$ so that opposite sides of the quadrilateral $PQRS$ are equal *and* parallel. Hence, $PQRS$ is a parallelogram.

Example 4.2.4. If 1800cm^2 of cardboard is available to make a box with a square base and an open top, find the dimensions of the box with largest possible volume.

Solution. Consider first, what we *want*, what we *know* and try to *connect* the two with information that enables us to calculate the unknown.

We *want*: to know the dimensions of the box that will lead to the greatest volume.

We *know*: that the dimensions of the box are constrained by the fact that only 1800cm^2 of material is available to make the box.

We begin by drawing a diagram and using suitable notation to translate what we want and what we know into mathematics. Let b be the length of the side of the (square) base of the box and h the height of the box (refer to Figure 4.2).

Figure 4.2

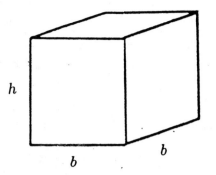

We *want*: b and h so that Volume is at a maximum

We *know*: The surface area of the box is to be 1800cm^2. Noting that the box has a square base and an open top, the surface area is composed of the sum of the areas of the base (b^2) and the four sides, each of which have area hb. Hence, the surface area S of the box is given by

$$S = b^2 + 4hb$$

This is to be 1800cm^2. Consequently, we can write

$$1800 = b^2 + 4hb \qquad (4.2.2)$$

To connect *what we want to what we know*, we note that the volume of the box is given by

$$V = b^2 h \qquad (4.2.3)$$

More precisely, from (4.2.2)

$$h = \frac{1800 - b^2}{4b}, \quad b > 0 \qquad (4.2.4)$$

Using this in (4.2.3) leads to

$$V = \frac{b^2(1800 - b^2)}{4b}, \quad b > 0$$

$$= 450b - \frac{b^3}{4}, \quad b > 0 \qquad (4.2.5)$$

This is now our mathematical model which will enable us to find what we *want*. The first step is to find the value of b which maximizes the expression (4.2.5) for V. Once this value of b has been found, we use (4.2.4) to find the corresponding value of h. We begin by noting that the domain of the function $V(b)$, given by (4.2.5), is $(0, \sqrt{1800}]$ (otherwise, $V < 0$). Since this interval is not closed, we cannot use the *"Extreme Value Theorem"*. However, since V is continuous on its domain, we can use the *"First Derivative Test for Absolute Extema"*. This requires that we first find all critical points of the function V. The set of critical points consists of those values of b obtained from equating $\dfrac{dV}{db}$ to zero and those values of b at which $\dfrac{dV}{db}$

does not exist. From (4.2.5), V and its derivatives are polynomials (and therefore continuous). Hence, there are no values of b at which $\dfrac{dV}{db}$ does not exist.

Consequently, all critical points of the function V arise from equating $\dfrac{dV}{db}$ to zero:

$$
\begin{aligned}
V(b) &= 450b - \frac{b^3}{4} \\
\frac{dV}{db} &= 450 - \frac{3b^2}{4}
\end{aligned}
\tag{4.2.6}
$$

Hence,

$$
\begin{aligned}
\frac{dV}{db} &= 0 \\
&\iff b = \pm 10\sqrt{6}
\end{aligned}
$$

We eliminate the value $b = -10\sqrt{6}$ since we require that $b > 0$. Alternatively, we can argue that the physical nature of the problem requires that we eliminate the (mathematical) solution $b = -10\sqrt{6}$ since the box cannot have a negative dimension (again, the mathematical model has given us *extra* information - we select that which is relevant using the information provided in the problem). It follows that V has only one critical point, $b = 10\sqrt{6}$. From (4.2.6),

$$
\begin{aligned}
\frac{dV}{db} &> 0, \quad b < 10\sqrt{6} \\
\frac{dV}{db} &< 0, \quad b > 10\sqrt{6}
\end{aligned}
$$

In other words, V is increasing for all b to the left of the critical point and V is decreasing for all b to the right of the critical point. From the *"First Derivative Test for Absolute Extema"*, it follows that V is at an *absolute maximum* when $b = 10\sqrt{6}$. From (4.2.4), the corresponding value of h is given by

$$
\begin{aligned}
h &= \frac{1800 - b^2}{4b} \\
&= \frac{1800 - (10\sqrt{6})^2}{4(10\sqrt{6})} \\
&= 5\sqrt{6}
\end{aligned}
$$

To summarize, the largest possible volume is given by

$$
\begin{aligned}
V_L &= V\left(10\sqrt{6}\right) \\
&= 3000\sqrt{6} \\
&\sim 7348.5 \text{ cm}^3
\end{aligned}
$$

occurring when

$$
h = 5\sqrt{6} \text{ cm} \qquad \text{and} \qquad b = 10\sqrt{6} \text{ cm}
$$

Chapter 5
HOW TO BE SUCCESSFUL IN EXAMINATIONS

Examinations invariably make up the single largest contribution
to a student's final grade in any mathematics course

This simple fact explains why the majority of students enrolled in mathematics courses are focused on course examinations (quizzes, midterms and finals) rather than on an appreciation of the value of course material. Simply put, your performance on course examinations will determine how well you do (overall) in your mathematics course. For this reason, it is absolutely essential to understand the *reasons why* the very best students are so successful in course examinations and how to use this information to your advantage.

So why do certain students perform so much better than others on course examinations? Some students put it down to a simple matter of intelligence: "Oh, that girl is really smart, her father is a physics teacher and her mother has a Ph.D.! No wonder she scores over 90% on all her tests!". Others put it down to the lack of a social life: "That guy never goes out. He does nothing but study. No wonder he performs so well on tests!". I suggest that neither is entirely correct and the truth lies somewhere in between. At the senior high school and junior college/university level, 'intelligence' alone is no longer sufficient to place someone in the top 5% of the class. There is far too much material to absorb and not enough time to do it in - even if you devote all your time to studying (in fact, it has been my experience that the very best students have an extremely active social life! - the level of activity increasing with the level of success!!). The answer has more to do with *how* you prepare for an examination and *what you do* to prepare.

I learned early on in my academic career that

knowing the course material   ⇎   success in course examinations

I recall my midterm examination from my first beginning calculus course. I had worked consistently throughout the year, understanding the course material, doing every assignment and working through extra practice problems (much in the same way as I did when I was in high school). I understood the main ideas and concepts and I was able to apply them in different situations, so I was quite confident that I would do well on the examination. Imagine my surprise when I discovered that I had scored only 58% ! Worse than that, many of those students scoring above me had performed poorly throughout the course - missing assignments and often asking *me* for help! I couldn't understand why. I had worked hard, I knew the material, so why wasn't I performing to the best of my ability? I began to discover the answer to my question when I asked one of my study-group partners (who had the highest score on the test) how she had prepared for the examination. It became clear to me that there were some *missing ingredients* in my test preparation routine. Basically it came down to two things:

- Smart-Practice

- Examination Technique

My friend and I had both prepared well *during* the course. What made the difference in our midterm scores was what we did in *preparing for the test itself* . She had obviously regarded the test as a *separate entity, targeting* and tailoring all her efforts not solely toward reviewing the course material (as I did) but to **doing well on the test itself** *(smart-practice)*. She had obtained many former and practice midterm examinations and *rehearsed* her performance so that she had a much better idea of what was *expected* and how to demonstrate the required knowledge under a time-constraint (*examination technique*). She was entirely focused on doing well *on the examination*. I, on the other hand, was focused *on the course material* believing that to be sufficient to perform to the best of my ability on the midterm.

To understand why my friend's strategy was so much more effective than mine, let us return to the *car-driving analogy.* None of us believe that we can pass a standard driving test simply by driving the way we do in 'everyday life'. We recognize that a driving test requires us to demonstrate a distinct collection of manoeuvres and exercises, *based* on basic driving skills, *under examination conditions.* Conversely, no-one continues to drive the way they did during their driving test! The latter is a rehearsed performance requiring specific targeted practice based on a knowledge of exact requirements (*smart-practice*) and a focused effort to perform well under specific test conditions (*examination technique*). Consequently, in preparing for a driving test, we find out as much as we can about *what* is required and *target* our preparation (as effectively as possible) toward those particular goals. Exactly the same principles apply to preparing for *any* test - academic or otherwise.

In fact, since that first calculus midterm, an acknowledgment of these basic principles has allowed me to perform to the best of my abilities on all subsequent examinations - academic examinations (in many different disciplines), driving tests, athletics competitions or whatever requires me to demonstrate performance under a given set of constraints.

In this chapter, we discuss, in detail, the many different aspects of maximizing performance on examinations, including, in particular, the two main ingredients: *smart-practice* and *examination technique*.

5.1 PREPARING FOR EXAMINATIONS: SMART-PRACTICE AND EXAMINATION TECHNIQUE

To prepare effectively for course examinations, it is important to think about *both* long- and short-term preparation. Each is an essential component of an effective overall exam-preparation strategy. Long-term preparation builds a *solid base or foundation* in the course material, allowing time for learning, understanding and asking questions. The material is slowly 'digested' and 'assimilated' into long-term memory where it can be recalled relatively simply (with, for example, a few practice problems) as required.

Short-term preparation is more concerned with the details of the examination itself and involves fine-tuning, targeted (or smart-) practice and practice in examination technique.

There are many examples in real-life where this type of long/short-term strategy is employed. For example, when training for competition, athletes build a solid base throughout the year using weight training and a combination of speed and endurance exercises (long-term preparation). Only as competition time approaches, do they tend to devote most of their training to their *specific* events (short-term preparation).

Long-Term Preparation

Clearly, what you do *during* term time will affect your performance in course examinations. As noted above, long-term preparation is not, in itself, sufficient to guarantee maximum performance in examinations but it is certainly *necessary*. Those who omit the long-term component of exam-preparation engage in what is more commonly known as *cramming*.

Cramming describes an effort to fit a significant amount of work (for example a month's or entire semester's worth of work) into a single night or weekend. **Cramming never works in mathematics** - basically because mathematics is not fact-based. Rather, mathematics is *method-based, cumulative* (the understanding of one part depends heavily on the understanding of previous parts) and *example-driven.* There is no way to read, understand and practice every technique from a mathematics course in a such a short period of time. It just doesn't work! - those who try cramming for a mathematics examination *never* perform to the best of their ability.

Fortunately, long-term preparation takes place automatically for those that work consistently throughout the course. The main components are summarized below for your convenience.

- **Lectures/Classroom time**

    - Collect information in a clear, concise and organized fashion. This will make it easier to understand when it comes to reviewing for examinations (see Section 3.1).

    - Pay particular attention to topics/sections of the course material and examples emphasized in class. They invariably show up on examinations!

    - Make notes of any hints or *extra information* that an instructor might give throughout the course.

    - Use class/lecture materials to identify relevant sets of practice problems from the textbook or otherwise. These will be useful when it comes to practicing or 'fine-tuning' for examinations.

- **Assignments and Homework**

    - Write clear, concise and logical solutions. This will help tremendously when it comes to reviewing for examinations - it is always easier to *review* than to *relearn* (see Section 3.3).

    - If you answered any assignment problems incorrectly, make sure you understand your mistakes and make a note of the correct solution(s). It is always best to do this *at the time* i.e. when the topic is current and fresh in your mind. When you return to these assignment solutions (for example, when reviewing), not only do you have the correct solution, you also have a note of what *not to do* i.e. your original error!

    - Use assignment questions as an indication of what's important and the standard expected by your instructor. I often tell my students to think of assignments as a collection of 'study-guides' for each particular topic covered in the course.

- **Developing Efficient Problem-Solving Techniques**

Using clear and logical procedures to solve a wide range of problems over the course of the semester cannot help but develop efficient problem-solving techniques. The latter improve only with experience and practice - something which cannot be achieved in the short-term. By examination time, many of these techniques will become *automatic*, requiring only *fine-tuning* when reviewing for examinations.

- **Asking for Help**

'Fix' problems as they arise (not all at once near exam-time - this leads to panic and exam-anxiety). Use *all* of the resources at your disposal, including your instructor (see Sections 3.4 and 3.5). Make a note of the 'help' you receive - you may need it later, when reviewing, if you encounter similar difficulties.

The most successful students continue to demonstrate the importance of consistent *long-term* effort as a necessary foundation on which effective exam-preparation strategies are based. There is no doubt that this has always been and will continue to be an extremely significant factor in distinguishing optimum from mediocre performance in course examinations.

Short-Term Preparation

Short-term preparation is most concerned with preparing for the examination itself and involves fine-tuning, *targeted* (or smart-) practice and practice in examination technique. The following are the main components of an effective short-term strategy.

- **Find Out What Will be Covered in the Examination**

It is always extremely useful to know which material is examinable - and which isn't ! This allows you to target your efforts toward the relevant areas. There are basically two ways to get this information.

(a) *From (recent) past examinations in the same course.* In many introductory mathematics courses, course examinations tend to cover the same standard material, year after year. For example, in typical beginning calculus or linear algebra courses, the midterm and final examination topics are almost always the same. Use this information to your advantage. Note any patterns, relative emphasis of one topic as opposed to another and, perhaps most important, which topics appear most often. See below for information on how and where to obtain old or practice examinations.

(b) *From your instructor.* ASK your instructor ! This is a perfectly valid question and you have nothing to lose (but remember to be polite, courteous and professional). In most cases, the instructor makes up the examinations so there is no better person to ask. Similarly, when examining old course examinations (as in (a) above), show them to your instructor. Ask if this is "a good example of what will be on the exam". Ask if he or she recommends that you work through any *specific* past examinations or set of practice problems.

- **Smart-Practice**

Once you have a good idea of the topics most likely to appear on the examination, concentrate and target your efforts toward these particular areas.

1. Review the relevant theory from course notes and/or the course textbook.

2. Review relevant worked examples from course notes and/or the course textbook.

3. Review your solved assignments.

4. WORK THROUGH RELEVANT PROBLEMS!

    (i) *Practice Problems.* Perform sets of relevant practice problems from the textbook or any other (preferably suggested - by, for example, your instructor) source. Choose these practice problems carefully: the main purpose here is to develop *fluency* and a *working knowledge* of selected techniques. Consequently, the practice problems should be mainly repetitive. Once you feel comfortable with any particular method, try a different style of problem using the same technique but perhaps in a different setting (these types of problem usually appear towards the end of a particular problem-set in the textbook). If you are unsure about whether or not a particular set of problems is relevant, ASK your instructor! A good rule-of-thumb is to perform at least five (standard) practice problems per technique (to develop fluency) followed by one or two 'different'/unusual problems in the same technique (for fine-tuning).

    (ii) *Re-work Relevant Assignment Questions.* This is excellent practice since you will have access to the correct solution with which to check your work. Also, assignment problems are good indicators of the required standard and are chosen specifically to reinforce relevant lecture (and probably examinable) material.

    (iii) *Suggested Problems.* Your instructor may have worked through or suggested specific problems in class while a particular topic was under discussion. Work these problems (again, if necessary) for they have been chosen specifically to reinforce or demonstrate a particular method or technique. Finally, *ask* your instructor to recommend some practice problems. You can be sure that this information will be relevant!

    (iv) *Work Through Old or Past Examinations.* This is discussed in detail below.

- **Work Through Old/Past Examinations.** There is no doubt that this is the most crucial stage of 'smart-practice'. An old or practice examination affords you the opportunity to actually *rehearse* the event. The questions are as *relevant* as they can be (they are *actual* examination questions for the very same math course !) and the examination conditions are probably identical (if not extremely similar) to those that you will encounter. There are two equally important components of an old or practice examination package:

(a) *The Examinations Themselves.* These can be obtained from many different sources:

   (i) Your instructor.

   (ii) Departmental offices.

   (iii) Bookstore.

   (iv) Exam Registries in the Student's Union or equivalent.

   (v) From students who have taken the course previously.

It is always a good idea to ask your instructor which old or practice examinations are most relevant and which he or she would recommend!

(b) *Detailed Solutions to the Examinations.* These are usually extremely difficult to obtain - but get them when you can! They are extremely valuable ! Not only do they allow you to check your solutions but they let you see what is expected and the required standard. Sometimes, instructors will make solutions available, sometimes they are sold in packages in departmental offices or bookstores. It may take a little effort to find them but it is always worthwhile - they make a particular examination much more effective as a tool for practice/rehearsal. Be careful, however, to use the solutions properly. Don't read them as a substitute for working through the problems!

   (Reading the solution $\nLeftrightarrow$ doing the problem yourself !)

   The solutions will always look easier than expected and there is no substitute for performing the procedure yourself. You should pretend you *don't have the solutions*, struggle with the problems as necessary (this is where the majority of the learning takes place) and consult the solutions only after you feel you have 'finished'. Remember, there will be no solutions available in the examination! If you cannot get solutions to a particular practice examination, try the examination yourself and then ask a member of the teaching staff to help you with any difficulties - or to check your solution technique for any obvious mistakes. Remember, the correct answer is only part of the solution - the technique by which you arrive at the answer is just as important (see Section 5.3).

There are two ways to use old/practice examinations:

1. *As an Excellent Source of Relevant Practice Problems.* Ignore the examination conditions and use the problems for practice. This will develop the required fluency and fine-tune your skills.

2. *As a Way to Develop Examination Technique.* When you take into account the actual examination conditions, practice examinations allow you to *rehearse* while actual old examinations allow for *dress-rehearsal.* You can actually simulate the examination itself! This will allow you to develop the necessary skills required to perform under the constraints (time or otherwise) of an actual examination (see below).

I always recommend to my students that they should work through at least three practice examinations before writing their particular examination. At least one of the three examinations should be an actual old/past examination - and used for *dress-rehearsal.*

• Examination Technique

Examination Technique is concerned with two things: *acknowledging* that an examination requires you to perform under certain constraints and *practicing*, as much as possible, to overcome any difficulties associated with these constraints (for example, exam-anxiety and time management). The constraining elements inherent in most examinations are:

- Examinations always incorporate an unknown element i.e. you are never 100% sure of what you may be asked.

- Examinations require that you perform under a time constraint.

- Examinations require that you demonstrate your knowledge precisely and logically. This means that, to be most effective, you must present your solution in a manner compatible with that expected by the person grading the examination. We shall discuss the different aspects of writing an examination, including the most efficient way to present solutions and what the grader looks for when grading, in Section 5.3.

Taken together, these constraints are largely responsible for the two most common complaints associated with writing examinations: *exam-anxiety* and *insufficient time.*

– *Exam-Anxiety.* It has been my experience that the most successful students overcome exam-anxiety by making the examination an *anti-climax.* By the time they get to the examination, they have worked through so many practice problems and practice examinations that they are on 'auto-pilot'. They know what to expect, many of their reactions during the examination are instinctive and they are *focused* on their particular goal. There will always be the usual 'adrenaline-rush' associated with writing an examination but, when it comes to overcoming exam-anxiety, *preparation* is the key, particularly short-term preparation, specifically *rehearsal* and *dress-rehearsal.* I have found that the process of working through practice examinations is one of the most significant components of overcoming exam-anxiety in mathematics.

– *Insufficient Time.* This is again symptomatic of inadequate practice and rehearsal with *actual* examinations. Working through an adequate number of practice problems allows you to develop fluency in the necessary skills making your approach and problem-solving procedure almost automatic. This means that, in the examination, you solve problems quickly and effectively. Practice/old examinations on the other hand, *tell* you what to expect in the allotted time. Working through a sufficient number of practice examinations cannot help but to inform you of how much material you are likely to encounter in the examination itself. You will know (approximately) what to expect and you can practice performing the required number of problems in the allotted time. We will return to this particular topic later in Section 5.3 when we discuss *writing the examination.*

It is clear that old/practice examinations play an important role in developing *Examination Technique.* There are other important factors determining your effectiveness when *writing* an examination. These will be discussed in Section 5.3.

5.2 PREPARING FOR EXAMINATIONS: GETTING ORGANIZED

As a mathematics instructor, I have always believed in going from the specific to the general. My approach to this particular chapter is no different. In Section 5.1 we discussed the specifics of how to study for a mathematics test: *smart-practice* and *examination technique.* In this section we take one step back and examine the more general issues related to exam-preparation.

How you prepare for an examination depends very much on the particular type of examination you are required to write. For this reason, it is important to obtain as much general information as possible about the examination itself, as soon as it becomes available:

- **Details of the Examination:**

  - Quiz, Midterm or Final ?

  - Place, Time and Duration ?

  - Open Book or Closed Book ?

  - Multiple-Choice, Written - or both ?

  - Are 'Cheat Sheets' (Formula Sheets) allowed ?

  - Are calculators allowed ?

  - How much is this exam worth as a percentage of the final grade ?

  - Absence Policy - what happens if you miss the exam for any reason ?

- **What's Examinable ?**

  - Once you have discovered the type of test facing you, you should then enter your *short-term* preparation routine as discussed above. The first step in this routine is to find out what is likely to be asked on the test and what isn't. This will begin the process of *smart-practice*, as discussed above.

We have already discussed, in detail (see Chapters 3 and 4), some of the things you should do as part of an effective long-term preparation strategy. The following are additional suggestions which may help you during your short-term preparation:

- **Prepare a Review Schedule**

Prepare a structured review schedule. This schedule will vary in length and depth depending on the volume of material you have chosen to study. This, in turn, will depend on the type of test you are required to write. For a final examination, prepare your schedule at least two weeks in advance; for a midterm examination, one to two weeks in advance whereas for a quiz, often a few days in advance will suffice. When preparing your schedule:

  - Overview the relevant course material and divide the material into separate sections - usually based on different theories, techniques or applications.

  - Identify the examinable material.

  - Allocate study time to each section according to the volume of material, its relevance to the examination and its importance (i.e. the likelihood that it will appear on the examination).

- For each section, identify a set of relevant practice problems from the textbook, course notes, past assignments or otherwise. Make a note of these.

- Decide which practice/past/old examinations you will work through and when (remember to keep one for the final stages of your review as a *dress-rehearsal*).

- Remember to allocate more time for particularly difficult concepts or examples which require more thought e.g. word problems (see Section 4.2).

- **Review Section by Section**

Review the material section by section as follows.

- Begin with an overview of the concept/theory or technique in a particular section.

- Read through worked examples in that technique (from e.g. course notes or textbook) until you feel confident enough to be able to apply the technique for yourself i.e. until you understand the main ideas behind the method.

- Begin practicing the method for yourself using the sets of (targeted) practice exercises identified earlier (above). Be sure to write clear, logical and *methodical* solutions to the problems (see Section 3.3 and Chapter 4), as if you were teaching the material to yourself - this will help you pick up extra points when writing the examination (see Section 5.3).

- Once you feel comfortable with the technique(s) in a particular section, move on to the corresponding practice problems from course notes and old assignments. You should have access to the solutions to these problems so make sure you check your *method* as well as your answer (the former is more important - see Section 5.3).

- By this stage, you should have a good working knowledge of the section of material under review. To 'fine-tune' your skills, pick a problem or two (dealing with this material) from some recent/practice examinations and see if you can confidently and competently write a *full* solution to each problem. DO NOT SKIP ANY STEPS - get into the habit *now* of writing fully comprehensive solutions - remember, in the examination, you must DEMONSTRATE your knowledge. Don't assume that any particular step is trivial - it may not be to the person grading the examination (see Section 5.3 for more on this)!

- Finally, to complete the review of a section, write a summary of the section as follows.

* List the important concepts/techniques/formulae in the section.
  * Link each concept/technique/formulae with practice problems, assignment problems, examples in course notes or the textbook and/or practice examination problems which you have worked through as part of your review and which you have found to be particularly good for understanding and developing fluency in the technique. You may want to return to these problems etc for a quick review of the section, nearer the examination.
  * Make any little notes which you think may help you when you return to this material later.

  − Follow this procedure for each section of examinable material.

- **Study Groups - Discussion**

Group review is particularly effective for the following reasons.

  − Active/Cooperative learning - it is always a good idea to *talk* about the material under review.

  − Students see and learn problem-solving strategies.

  − Students learn from other students.

  − Individual students may get stuck but the group will tend to keep the momentum going.

  − Improvement in self-esteem and a decrease in anxiety levels.

Consequently, IN ADDITION TO (not instead of) the above review procedures, make an effort to discuss the material/examples/problems with other people. Be careful to use this activity properly (as a *supplement* to your review). I have witnessed many situations where students working in groups believed they were reviewing effectively but were, in fact, merely 'taking note' of the efforts of the well-organized students who always led the discussions (see the discussion on group work in Section 3.3). When all is said and done, you will face the examination alone. It pays to keep this at the back of your mind - at all times!

- **Final Review and Fine-Tuning**

Following the above procedure allows you to learn the material, develop *fluency* and *method* in the particular techniques and commit all of this to a part of your brain from which it can be easily retrieved. In fact, in the run-up to the examination (e.g. one or two days before) review the collection of summaries made when reviewing each individual section. Build confidence and reinforce what you already know by selecting random practice problems from your collection and writing full 'blind' (i.e. without any supplementary information such as notes or textbook - just as you will

in the examination) solutions to each problem. Finally, fine-tune your skills by
rehearsing with one or two complete examinations. After this, you will be
well-prepared for anything!

Finally in this section, we make a few 'common-sense' suggestions relating to
the logistics of studying.

- **Where to Study**

To study mathematics effectively, it is necessary to free yourself of distractions and
competing associations. I have found over and over again that the mathematical
part of the brain really 'warms up' only after a period of deep thought or effort,
usually after struggling or trying very hard to solve a particular problem.
For this reason, you should *never mix business with pleasure* when it comes to
studying for examinations. Instead, you should try to make your study time as
efficient and as effective as possible. In my experience, one hour of concentrated,
focused study is worth three of 'watered-down', distracted study-time any day! Try
to keep the following simple rules in mind:

 - Pick a quiet room free of distractions. For example,

   * Bedroom
   * Library

NOT

   * Your bed
   * The living room of your home
   * The kitchen table
   * A local fast-food restaurant
   * A cafeteria or coffee shop

- **Get Comfortable**

Make sure that when you are studying, you wear comfortable clothes and use a
comfortable chair. Any discomfort will distract you from your main purpose.

- **Take Frequent Study-Breaks**

Be sure to get out of your study environment for a study-break whenever you feel
the need - perhaps every hour or so. This will keep you sharp and maximize your
effectiveness. Beware however, study-breaks shouldn't occur too frequently -
remember, it usually takes at least 30 minutes of effort to *warm-up* your thinking.
You shouldn't interrupt your concentration just when you 'get into it'. Breaking too
frequently will mean that you are constantly warming-up and never working at the

most effective level. Keep the breaks short (e.g. 5 minutes) and simple - get up for a stretch, a snack or something - but be careful to minimize distractions during your break - keep things rolling over in your mind and don't get into some 'deep' (unrelated) conversation with a friend!

- **Eat Well and Get Plenty of Rest**

To perform well mentally, you need to stay healthy. If you organize your study time effectively, you will have sufficient time to eat well and get lots of sleep.

- **Physical Activity**

For some reason, after spending a significant period of time thinking about a problem, things often 'come to me' in the most unusual places or when I'm doing something completely unrelated e.g. while I am jogging or working out at the local gym. I'm not sure why but the complete change in activity (from studying) seems to make things clearer. Many of my friends and colleagues have indicated similar experiences. The same may work for you. Even if it doesn't, some good physical activity does tend to refresh and re-energize both our minds and our bodies - and it doesn't have to be anything sophisticated! Even taking a brisk walk seems to have the desired effect.

5.3 WRITING THE EXAMINATION

The day has arrived! You're well-prepared, confident and ready to go. Nevertheless, to maximize your performance *during* the examination, there are certain essential components of exam-writing of which you must be aware.

Eat Something. Make sure you have a good meal on the day of the examination - you will expend lots of energy when writing the examination so make sure you fuel your body sufficiently.

Dress Comfortably. Wear comfortable clothes! Remember, you may be sitting (in the same position) for up to three hours!

Do You Have Everything? Before you leave for the examination room, go through a checklist of all the things you need to write the examination. For example,

- Writing instruments
    - Pens
    - Pencils
    - Rulers
    - Erasers

- Calculator (according to examination regulations)- remember to check the batteries - bring spares if necessary.

- Textbook or other supplementary materials allowed by the examiner (if open book).

- A watch.

Bring replacements/'back-ups' as necessary.

Get There Early. Arrive at the examination site at least 15 minutes before the examination begins. This will give you time to compose yourself, note the seating plan and make yourself aware of any (new) instructions.

Get a Good Seat. When you enter the examination room, make sure you choose a good seat - one which is relatively free of distraction. In some cases, large examination rooms are used for a variety of different examinations, of varying styles and duration. If your examination is two hours long, and the row next to yours is being used for a one hour examination, you will be distracted by students packing up mid-way through your examination. In such cases, it always pays to take a few minutes before the examination to study the seating plan. Similarly, try not to sit near students with heavy colds - they tend to sniff and cough a lot! Students (even friends) with a different examination philosophy (i.e. those that tend not to take examinations seriously) should also be avoided when it comes to seating. You have invested too much time and energy to risk being distracted by someone who is not focused. Find a nice quiet area, not too far from the people proctoring the examination - you may need to ask questions!

The large majority of examinations in mathematics are of the written type, multiple-choice or an element of each. We begin with a discussion of the written examination and return to the subject of multiple-choice tests later in the chapter.

Maximizing Performance in Written Examinations

The First Few Minutes. When the examination begins, spend the first few minutes scanning the questions (including the mark distribution). When doing so, note (in writing - near each question) which technique you will use and how much time you think you will need to answer each question. These little notes will help you allocate your time effectively and act like 'doors', opening compartments to the (now) more than familiar corresponding review sections. Scanning the entire test at the start also gives you an opportunity to make sure that your examination paper is complete - imagine, with 5 minutes to go, finding out that you are missing a question worth 20% of the grade! Once you have looked at all the questions, rank them according to level of difficulty. The 'easy' problems are usually of TYPE A (Section 4.1). These problems require you to demonstrate method and set procedures. The TYPE B problems (Section 4.2) are often classified as the most difficult, requiring more thought and less routine application of course material.

At this point you can proceed in one of two ways depending on your particular preference:

1. *Start with the easiest problems and work towards the most difficult problems.* There are three main advantages to this strategy:

   (i) In solving the easier problems, you slowly *warm-up* your thinking in preparation for the more difficult questions.

   (ii) Solving a series of problems successfully means that you gain immediate confidence for tackling the more challenging problems.

   (iii) Getting the easier problems out of the way first, will maximize the amount of time remaining to consider the more challenging problems.

2. *Start with the most difficult problem and work towards the 'easiest' problem.* The main advantage in using this approach is that you can tackle the questions requiring the most thought (TYPE B) at the beginning of the examination when you are less tired and more alert.

In my own mathematics courses (those in which I have been the instructor)), the majority of the top students have consistently favored the first approach (which is also the method I used as a student) - but it depends on your particular preference.
Once things are underway, bear in mind the following - they will add to your overall effectiveness when writing the examination.

Note the Mark Distribution. Use the mark distribution to allocate time for each question. Clearly, a question worth 5 marks should not require as much time as one worth 20 marks. Use the mark distribution also as an indication of what is expected i.e. the '20 mark-question' requires '20 marks worth of effort' etc.

Detail in Open Book Exams. Open book examinations rely less on memory and more on method and technique. Consequently, you are expected to demonstrate more detail in open book examinations than in closed book examinations. For example, in a closed book examination you may get a mark or two for writing down a correct formula. In a closed book examination, this is considered as being 'supplied' (in the text) and so carries no weight. Instead, open book examinations emphasize more method and problem-solving techniques.

Formula Sheets. If a 'formula sheet' is supplied with your examination, use it as a guide to the techniques to be used in the examination. For example, if a complicated formula does not appear in the formula sheet, it is unlikely that you will need to use that formula.

Try Every Question. Don't be afraid to try to answer every question-even if you're not sure how to proceed. Write down your thoughts and try to develop a solution using logical steps. Part-marks may be awarded for some of the things you write down.

Watch Your Time. Pace yourself and try to stick as closely as possible to the time you have allocated to each question (on your initial appraisal of the examination). If you get 'stuck' and can't seem to make any progress on a particular problem within the allocated time, LEAVE IT and plan to return to it at the end of the examination. It is better to lose marks on one question and gain marks on the remaining problems than to sit for the rest of the examination wasting valuable time. Remember, even if you have been engaged in active thought when answering a particular question, if there is nothing on paper, the instructor will assume that you have done nothing in that time! There is no way for the instructor to believe otherwise.

Ask for Clarification. If you are unsure about anything to do with the writing of the examination, including the wording or the statement of a particular problem, ASK! There is nothing to lose and you gain the added advantage that a verbal clarification might 'jog your memory'.

Write for Maximum Points. This is perhaps the most important aspect of writing examinations. An *examination* is just that! An *examination/investigation* of a students performance, in a particular subject, on a particular day. As such, a student is required to *perform* i.e. to *demonstrate* his or her knowledge of the subject. When you bear in mind that in a written examination, the only way to *demonstrate* ability or knowledge is by writing it down, you will understand that the person grading the examination will use the *written response* as the sole criterion to judge a student's ability to answer a particular question. This is *the* crucial consideration which *must* be taken into account when writing examination solutions. Solutions must be written with the person grading the examination in mind.

After I give an examination, students always return to 'discuss' their performance. Some wish to see their examination paper in an effort to 'pick up' more marks. What follows are some of the explanations I have heard whenever students have tried to explain 'blank or partial solutions' to examination problems.

- "I knew what I was doing, I just didn't write it down!"

- "You" (meaning the instructor) "know that I know how to do this stuff! I've done it many times in the assignments - and you gave me 100% each time. Surely you didn't expect me to produce all the detail. There is a time limit you know!"

- "I worked out the problem on a piece of scrap paper. I didn't write the details down because I didn't think they were important. I did get the correct answer, however - look!" (student points to the one equation $x = 3$ on the otherwise blank page). "Don't I deserve full marks for this question?"

- "I didn't have time to write the complete solution so I did the calculation quickly in my head and obtained the correct answer. Why did you give me only 1 mark out of 10 ?"

I answer these 'comments' by informing the students (again! - I do this in class at least 3 or 4 times *before* the examination) of exactly what I (and mathematics instructors, in general) look for when grading examinations:

- A set procedure/method illustrating clear, logical thinking and understanding, leading to the correct answer.

- An ability to use the most appropriate technique in the most efficient way possible.

- An ability to communicate ideas using the language of mathematics.

- An ability to develop a mathematical argument, stage-by-stage, step-by-step, leading to the desired result.

- An ability to use problem-solving strategies and explain the significance of any results obtained.

Clearly, the final answer (e.g. $x = 3$) is only one part of the *solution* - accordingly, it carries only a proportion of the points allocated to a *complete* solution.

Earlier, in Section 3.3, we discussed the importance of writing clear, logical, step-by-step solutions to assignment problems. *Exactly the same is true* when writing solutions to examination problems - except that in this case, it is even more crucial! Your instructor will equate what you have on paper to what you know about solving a particular problem - there is nothing else!

Figures 5.1 and 5.2 contain two different (graded) solutions to a problem from a final examination in 'Precalculus'. Figure 5.3 contains the instructor's solution and the marking scheme.

Figure 5.1: Student #1

(10 points) 1. (a) Solve the inequality

$$|3x - 1| - |x + 1| > 0$$

$|3x - 1| >$

①

$3x - 1 - x - 1$

$|3x - 1| > |x + 1|$
$9x^2 - 6x + 1$
$> x^2 + 2x + 1$
$9x - x^2$
$8x^2$

① $x < 0$ or $x > 1$

(b) Find the domain of the function

$$s(y) = \frac{1}{\sqrt{y^2 - 1}}$$

② $\sqrt{y^2 - 1} \neq 0$

$y^2 - 1 > 0$

$y^2 > 1$

$\dfrac{}{y}$

$|y|$

① $y < -1, \; y > 1$

$\boxed{\dfrac{5}{10}}$

59

Figure 5.2: Student #2

(10 points) 1. (a) Solve the inequality

$$|3x - 1| - |x + 1| > 0$$

$$|3x - 1| > |x + 1|$$

②

$$8x^2 - 8x > 0$$

$$8x(x - 1) > 0$$

①

$$\underline{x < 0 \quad \text{or} \quad x > 1}$$

(b) Find the domain of the function

$$s(y) = \frac{1}{\sqrt{y^2 - 1}}$$

$\boxed{\dfrac{5}{10}}$

①

$$y^2 - 1 > 0$$

①

$$\underline{y < -1 \quad \text{OR} \quad y > 1}$$

Figure 5.3: Instructor

(10 points) 1. (a) Solve the inequality

$$|3x - 1| - |x + 1| > 0$$

We require values of x such that $|3x - 1| > |x + 1|$. Since both $|3x - 1|$ and $|x + 1|$ are positive,

②

$$
\begin{aligned}
|3x - 1| &> |x + 1| \\
\Leftrightarrow |3x - 1|^2 &> |x + 1|^2 \\
\Leftrightarrow (3x - 1)^2 &> (x + 1)^2 \\
\Leftrightarrow 9x^2 - 6x + 1 &> x^2 + 2x + 1 \\
\Leftrightarrow 8x^2 - 8x &> 0 \\
\Leftrightarrow 8x(x - 1) &> 0
\end{aligned}
$$

②

$$
\begin{array}{ccc}
x<0 & 0<x<1 & x>1 \\
\hline
+ & 0 \quad - \quad 1 & + \\
\end{array}
$$

①

$$\Leftrightarrow x < 0 \text{ or } x > 1 \qquad ans$$

(b) Find the domain of the function

$$s(y) = \frac{1}{\sqrt{y^2 - 1}}$$

The domain of the function consists of all values of y such that $s(y)$ 'makes sense'. The expression $\dfrac{1}{\sqrt{y^2 - 1}}$ makes sense whenever values of y are chosen such that there is no division by zero and no square roots of negative numbers. Hence we require y such that

③

$$
\begin{aligned}
y^2 - 1 &> 0 \\
(y - 1)(y + 1) &> 0
\end{aligned}
$$

$$
\begin{array}{ccc}
y<-1 & -1<y<1 & y>1 \\
\hline
+ & -1 \quad - \quad 1 & + \\
\end{array}
$$

①

$$\text{i.e. } y < -1 \text{ or } y > 1$$

Hence, the domain of the function $s(y)$ is

①

$$y \in \mathfrak{R} \text{ such that } y < -1 \text{ or } y > 1 \qquad ans$$

61

Notice that, although both Student #1 and Student #2 arrive at the correct answers (Figures 5.1 and 5.2, respectively), neither demonstrates the (complete) correct logical procedures required to arrive at these answers (illustrated in the instructor's solution, Figure 5.3). Consequently, both score only 5 points out of a possible 10 - even though both obtained the correct answers! They may have *known* the correct procedures, they may even have *thought* through the problem using these procedures but the point is that they did not *write* down the details. The grader cannot and is not expected to 'read between the lines' or 'work out' what you are trying to say. Only what you *write down* will be used towards assessing your grade. Solutions are almost always graded on a *part-mark* basis i.e. marks are allocated to different parts of the *procedure*, of which the correct answer is but one! To get a perfect score, you must demonstrate *all* of the components of the procedure. This procedure also works 'for' you when, for example, you demonstrate the correct procedure but arrive at the wrong answer through an incorrectly performed calculation or otherwise. In this case, you will receive all the marks except those allocated to the correct answer itself.

Try to follow the following guidelines when writing solutions in examinations:

1. Recognize what the instructor will look for when grading your solution (see above).

2. Develop a clear, logical procedure leading to the correct answer.

3. Don't be messy! The grader will not take the time to 'decipher' what you are trying to say.

4. Don't assume that the grader 'knows what you are talking about' - tell the grader exactly what you want to say.

5. Label diagrams and place them where they belong - near the part of the solution to which they are most relevant.

Finally,

Don't Leave Early. Devote the *entire* examination to maximizing your performance! If you finish the examination early, use the extra time to check your solutions. Add detail, tidy up explanations, think about alternative strategies for questions you couldn't solve. Even if you think you have 'aced' the test, stay and check anyway.

Ignore Everyone Else. Sometimes, when fellow students leave the examination early, there is a tendency to think that the examination should be easy and that you are 'missing something' - nonsense! Take all the time you need. For all you know, those people leaving early may have failed. In fact, in my experience, examinations submitted ahead of time almost never account for the top scores in the class. Quite the contrary, the very best students use every available minute to their advantage.

Maximizing Performance in Multiple-Choice Tests

Most of what has been said above for 'written tests' is true also for multiple-choice tests. After all, both types of test have the same objective: to test the student's knowledge of the course material. From the student's point of view, the main difference and the biggest drawback to a multiple-choice test, however, is that, since each question requires only a simple *answer* e.g. A, B, C or D, there is little opportunity to *demonstrate* method, technique and understanding through the presentation of well-written, clear and logical *solutions*. Consequently, unlike written tests where method and procedure contribute significantly to the number of marks awarded for each solution, multiple-choice tests exclude the opportunity to accumulate marks from a detailed and methodical solution.

Nevertheless, in mathematics, the approach to solving problems on a multiple-choice test should, essentially, be the same as that used in solving problems on a written test. The main difference being that, at the end of the procedure, only the final answer is returned, not the solution (the *method* leading to the final answer). In other words, in a multiple-choice test, you will be answering the same types of questions as you would on a written test (only shorter - a consequence of asking many equally-weighted questions in a relatively short period of time). Consequently, you should be prepared to work through each problem, **anyway**, on a separate sheet of paper and *arrive* at the correct answer through the same logical reasoning used to solve problems on a written test. Accordingly, make sure you bring a supply of 'scrap paper'!

Of course, in a multiple-choice test, since the correct answer appears alongside each question (although 'hidden among decoys'), there is always the opportunity to *guess*. This 'method', however, should be used only as a last resort. **By far the most efficient and effective way of tackling a multiple-choice test in mathematics is to treat it, essentially, as a written test and return only the final answers (at the end of your solution) on the answer sheet.**

The following procedure may help you when answering multiple-choice questions.

Step 1. Identify any key words to help get through the 'padding'(extraneous information) and to the real meaning of the question. Identify the appropriate technique/theory to be used in the solution.

Step 2. Ignore the (given) answers and solve the problem for yourself on a separate sheet of paper. Do so clearly and methodically - as if you were supplying the solution to the same problem in a written test. If necessary, draw a diagram - it may help you to 'see things' more clearly.

Step 3. Compare your answers with the given alternatives. If your answer is among them, enter the correct choice on the answer sheet. If your answer is not among the alternatives but 'close' to one, go back and check your working (this will be easy to do if you have solved the problem methodically). Locate and correct the error and repeat *Step 3* until the correct alternative is identified. If, despite your best efforts, you cannot obtain one of the given alternatives, choose the one which is 'closest' to your answer.

Step 4. Move on to the next question.

The following tips may prove useful when answering multiple-choice questions.

- If you are completely 'stumped' and unable to even begin the solution to a particular problem, eliminate any alternatives which you *know* to be incorrect and make an *informed guess* from the remaining alternatives.

- The solution to a multiple-choice question is usually short and should take no more than 5 to 10 minutes (a consequence of asking many equally-weighted questions in a relatively short period of time). If you find yourself spending more than this amount of time on a particular question, you are probably using the wrong method - either guess or move on to the next question. Remember, each problem is worth the same number of marks!

- If you complete the test with time remaining, return to those questions which you answered with any degree of uncertainty. Re-examine the 'logical alternatives' (i.e. exclude the alternatives previously identified as 'known to be incorrect') and check your answers.

- Often, *working backwards* is a good way to check your answer. For example, suppose you are asked to find the integral

$$\int \sin x \cos x \, dx$$

and given the following alternatives:

A. $\cos x \sin x$

B. $-\dfrac{1}{4} \cos 2x + C,$ where C is an arbitrary constant.

C. $-\dfrac{1}{4} \sin 2x + C$

D. $\tan x - \cos x + C$

You should note that the correct answer will differentiate to the integrand $\sin x \cos x$:

$$
\begin{aligned}
\frac{d}{dx}\left(-\frac{1}{4}\cos 2x + C\right) &= \frac{d}{dx}\left(-\frac{1}{4}\cos 2x\right) + \frac{d}{dx}(C) \\
&= -\frac{1}{4}(2)(-\sin 2x) + 0 \\
&= \frac{1}{2}\sin 2x \\
&= \frac{1}{2}(2)\sin x \cos x \\
&= \sin x \cos x
\end{aligned}
$$

Hence, the correct answer is B.

5.4 AFTER THE EXAMINATION

After the examination, you should take any opportunity to view your graded examination paper. For class quizzes and midterms, this is usually done in class, on the day the instructor returns the graded scripts. For final examinations, it may be necessary to make an appointment with your instructor. No matter how well you did, reviewing your examination paper is a *learning experience and a vitally important exercise* in preparing you for the next examination - in the same course or otherwise. The following are some of the more important advantages associated with reviewing your examination paper.

Identify Errors in Grading. The instructor may have made an arithmetic error when adding the different marks awarded to each problem. Check the addition yourself. Similarly, the instructor may have 'missed a question' or forgotten to grade part of a problem - point this out. Ask the instructor to outline his or her grading scheme. This way, you can ascertain the various (part-) marks allocated to each solution and ensure that your paper was graded accordingly. Remember to be professional and courteous at all times.

Learn From Your Mistakes. Use the *examination experience* to improve your performance in any future examinations i.e. while the material is fresh in your mind, check posted solutions, try to understand any errors in your solutions and rework problems as necessary. This is particularly important when reviewing quizzes and midterms - you have the opportunity to correct any errors in method, understanding and presentation (writing) *before* the all-important final examination. Consult with your instructor as necessary.

Self-Critique. An examination affords you the opportunity to 'try-out' your studying, preparation and examination-writing strategies. Use your 'examination experience' to analyze, improve, develop and fine-tune your approach. Look for weaknesses and decide how you will not 'make the same mistake twice'.

Learn From the Examination. In the case of a quiz or midterm, you now have an actual examination made up by the very instructor that will be responsible (entirely or in part) for your final examination. Note the style, the emphasis and, above all, **what the instructor requires in a perfect solution.** Be sure to use this information to your advantage in the final examination. Quizzes and midterms also afford you the opportunity to have your instructor critique/comment on all aspects of your examination performance, **before** the final examination - what better way to improve your studying, preparation and examination-writing strategies than to have the instructor comment on the *results* of your efforts!

What to do if You Failed Miserably

If you really made a mess of the examination and you're not sure why, apart from doing the things mentioned above, you should make an appointment with your instructor and try to explain/discuss what went wrong. The instructor may see errors in your studying, preparation and examination-writing strategies that may be easy to correct. I make a point of meeting with students who find themselves in this position - and find the same recurring reasons why they have performed so badly on examinations (most of them have already been discussed in this book):

- Most admit to never having worked through a past or practice examination *by themselves.*

- Most have never engaged in 'smart-practice' (see Section 5.1).

- Most have never thought about 'rehearsing' and practicing examination technique.

In most cases, it is the **short-term preparation** which is inadequate or, in some cases, missing entirely. This is, however, easily corrected using the procedures outlined in Section 5.1.

It is important that you maintain adequate communication with your instructor, in this and in all course-related matters. At the end of the course, your instructor is charged with the responsibility of assessing your performance using the usual sources of information (assignment and examination scores) and any other information that *you care to make available* - you can make a good impression on your instructor by asking sensible (and frequent) questions in a courteous and professional manner. In this way, you appear conscientious and committed. Consulting with your instructor after an examination adds to that impression - and works to your advantage. For example, in some cases, instructors will discard a student's (disastrous) midterm and move the appropriate weight to the final examination. This usually happens only when the instructor is convinced that the student has the potential to demonstrate the required knowledge by the end of the course but, for example, has "got off to a bad start". Such a decision is almost always 'ad hoc' - based on the instructor's *impression* of the individual student's *demonstrated* abilities - a significant part of which depends on good communication between the student and the instructor.

Chapter 6
OTHER RESOURCES

Some students choose to seek help *outside* their mathematics course - mainly from tutors or from self-study manuals or study guides. Whereas there are many advantages to using any of the latter, the biggest drawback, by far, is the additional cost. For this reason, it is always best to utilize, as much as possible, the sources of help available as *part* of your mathematics course (see Chapter 3) **before** considering other sources.

In this chapter, we discuss the various different aspects of using a tutor or buying supplementary study materials. In particular, we shall be concerned with maximizing effectiveness and value for money.

6.1 USING A TUTOR

The main reason for hiring a tutor should be to obtain *specific, targeted, intensive, one-on-one help*, at convenient, selected times during a mathematics course. The additional cost involved means that most people hire tutors only when absolutely necessary - for example, at critical times during the course, whenever an extra injection of intensive help is required (e.g. when preparing for examinations).

Choosing a tutor, however, can be fraught with difficulties. Let's begin by noting a few important facts concerning tutors.

A tutor is **not:**

- Someone who "sort of" remembers something about the material or was "once able" to do something similar.

- A student who has recently passed the same mathematics course - even if he or she was at the top of the class! Passing a course (even with a perfect score) does not qualify someone to teach the material!

A tutor **is:**

- Someone with considerable expertise in the specific area of mathematics in which help is required. A general rule of thumb is that the tutor should have completed at least three levels of mathematics above that with which he or she is required to assist.

- Someone who can communicate ideas, in mathematics, easily and effectively.

- Someone who can answer specific questions quickly, efficiently and without any uncertainty.

The best way to find a tutor is through *word of mouth referral* - usually from your *instructor, a fellow student or an informed advisor.* Some institutions claim to have lists of 'approved' tutors. Be careful here! Make sure you ascertain what qualifies a tutor for inclusion on the list - sometimes it's nothing more than a simple phone call!

Once you have found a suitable tutor, it is necessary to take steps to ensure the most effective use of (*paid!*) time spent in consultation. In this regard, the following may help you:

Prepare! Before you meet with the tutor, make a list of all the questions you need to have answered. Keep the questions brief and specific. Make sure you are familiar with the material behind each question and the reason why you have asked a particular question. This will ensure that you make the time spent in consultation as effective as possible.

Control the Session. The tutor is there *for you.* Consequently, run the session according to *your* requirements. In this respect, make sure you tell the tutor exactly what you want. Stick, closely, to your list of questions. Don't let the tutor 'stray' from the topic and don't waste time.

Stay Focused. Every minute of time spent with a tutor costs money. Hence, stay alert and focused on your requirements. Don't leave a particular point or question until you are entirely satisfied with the explanation.

Targeted Help. Use the tutor as a *source of help* - **not to replace** your efforts on assignments and practice problems. The tutor should *add to*, not replace, the overall learning experience. The tutor cannot and should not do the work for you but rather *show you how.*

Groups. If you require less in the way of one-on-one attention but would like to have access to a tutor for the purpose of discussing a certain number of specific points and questions (e.g. while reviewing and/or working through old examinations), consider getting together with a few classmates and hiring a tutor *for the group.* This will save you money and afford you the advantages of *group discussion.* (see Section 3.3).

In conclusion:

- Choose the tutor wisely. Use word-of-mouth referral.

69

- Meet with a tutor only when necessary and only for as long as required (Bear in mind that most tutors insist on a minimum of one hour).

- Control the content and style of the meetings by being well-prepared and focused.

- Use the tutor to *add to*, not replace, the learning experience.

- Using a *group tutor* will save money.

6.2 USING SELF-STUDY MANUALS

Supplementary materials such as self-study manuals are similar to tutors in that they can be used to provide *specific, targeted help in relevant areas.* Consequently, the choice of self-study manual should be dictated by your specific requirements. For example, if you require a collection of worked past examination papers in Introductory Calculus, a book entitled *Understanding Mathematics* is probably less useful to you than one entitled *Calculus* which itself is probably less useful to you than one entitled *Introductory Calculus* which itself is probably less useful to you than one entitled *Solved Practice Examinations in Introductory Calculus.* The key to making the most effective purchase lies in the *relevance* of the material *to your requirements.* Targeted information is quick and to the point - and usually less expensive than that contained in books addressing more general issues.

When choosing a self-study manual/study aid, bear in mind the following:

- Choose a book which offers information closest to your requirements.

- Choose something which is relatively easy to read and which explains things clearly.

- Your instructor or someone with experience in the area may be able to recommend a book which meets your specific requirements.

- Word-of-mouth referral is extremely effective - i.e. ask what other people are using/have used/would recommend.

- One of the biggest advantages of buying information in 'book form' is that the information can be accessed almost anywhere, anytime e.g. in the bath, on the bus, in the library, in class or at home.

SUMMARY OF EFFECTIVE STUDY STRATEGIES

Before the Course Begins

- *Do you know what your instructor assumes you know ?*

Ensure you and your instructor start on a *level-playing field* by:

- Identifying necessary (assumed) prerequisite skills.
- Practicing necessary prerequisite skills (5 examples per 'skill'):
  * To achieve *fluency*.
  * To *warm-up* your thinking.

During the Course

- *Make the most of class time by:*

  - Gathering information neatly and concisely.
  - Using the lecture/class material as a guide to what's relevant.

- *Use your textbook:*

  - To reinforce lecture material.
  - As a source of illustrative examples.
  - As a source of practice problems.

- *Do all the assignments:*

  - To get that 10 - 20% of the course grade allocated to assignments.
  - To get practice in relevant techniques.
  - To help review for upcoming examinations.
    * Assignments identify problems which best illustrate key concepts.
    * Assignments tell you where to concentrate attention/practice.
    * Assignments identify problems similar to those likely to appear on examinations - the instructor is giving clues!

- *Be effective when writing solutions to problems (assignment, practice or otherwise):*

    - Write clear, methodical, *teach-yourself* solutions - so that later you need only review not relearn.

    - Organize your thoughts and develop effective problem-solving techniques.

    - Pick-up extra points from *part-marking* of a methodical, well-written solution.

    - In mathematics the solution (the steps leading to the answer) is more important than the final answer.

- *Use posted solutions*

    - To check your problem-solving techniques.

    - To see what constitutes a proper/correct solution (procedure) in the eyes of the instructor (the person who grades the examinations!).

    - To make sure that you are doing what you should be doing.

- *Engage in group study but ultimately be able to solve problems by yourself.*

- *Get help:*

    - From your instructor.

    - From any other available (as part of the course) source.
        * The key to learning mathematics is seeing examples and doing exercises - a large part of which is the ability to ask questions - *Ask! Ask! Ask!* - but do so politely and professionally.
            · Use the information in a *feedback loop* to improve your understanding and problem-solving strategies.

- *Develop effective problem-solving strategies by:*

    - Classifying problems according to type or class.

    - Identifying appropriate *keywords and phrases.*

    - Identifying the appropriate technique(s) to be used for each class.

    - Applying the appropriate techniques.

    - Practice! Practice! Practice!

Being Successful in Examinations

Preparation

- *Long-term Preparation*

    - Work consistently and effectively during the semester.

- *Short-term preparation*

    - Smart-Practice and Examination Technique
        * Target/focus your efforts appropriately.
        * Work through old/practice examinations - *rehearsal and dress-rehearsal.*

Writing Examinations

- *Note the Mark Distribution*

- *Pace yourself.*

- *Write for maximum points.*

    - Be clear, concise and methodical.

- *Be aware of what you are telling the grader.*

    - Don't assume that the grader knows what you are trying to say - spell it out!

Post-Mortem

Use the experience (examination or otherwise) to improve your learning and study strategies.

INDEX

Other titles in the *Smart Practices* series:

Calculus Solutions: How to Succeed in Calculus
0 13 287475 X

Coming in November 1997:

Ordinary Differential Equations
0 13 907338 8

Coming in 1998:

Statistics

Linear Algebra

Look for these and other *Smart Practices* titles in your college or university bookstore
or speak with a Prentice Hall Canada customer service representative at:

1 (800) 567-3800

NOTES:

NOTES:

NOTES: